Science and Fiction

Science and Fiction – A Springer Series

This collection of entertaining and thought-provoking books will appeal equally to science buffs, scientists and science-fiction fans. It was born out of the recognition that scientific discovery and the creation of plausible fictional scenarios are often two sides of the same coin. Each relies on an understanding of the way the world works, coupled with the imaginative ability to invent new or alternative explanations—and even other worlds. Authored by practicing scientists as well as writers of hard science fiction, these books explore and exploit the borderlands between accepted science and its fictional counterpart. Uncovering mutual influences, promoting fruitful interaction, narrating and analyzing fictional scenarios, together they serve as a reaction vessel for inspired new ideas in science, technology, and beyond.

Whether fiction, fact, or forever undecidable: the Springer Series "Science and Fiction" intends to go where no one has gone before!

Its largely non-technical books take several different approaches. Journey with their authors as they

- Indulge in science speculation – describing intriguing, plausible yet unproven ideas;
- Exploit science fiction for educational purposes and as a means of promoting critical thinking;
- Explore the interplay of science and science fiction – throughout the history of the genre and looking ahead;
- Delve into related topics including, but not limited to: science as a creative process, the limits of science, interplay of literature and knowledge;
- Tell fictional short stories built around well-defined scientific ideas, with a supplement summarizing the science underlying the plot.

Readers can look forward to a broad range of topics, as intriguing as they are important. Here just a few by way of illustration:

- Time travel, superluminal travel, wormholes, teleportation
- Extraterrestrial intelligence and alien civilizations
- Artificial intelligence, planetary brains, the universe as a computer, simulated worlds
- Non-anthropocentric viewpoints
- Synthetic biology, genetic engineering, developing nanotechnologies
- Eco/infrastructure/meteorite-impact disaster scenarios
- Future scenarios, transhumanism, posthumanism, intelligence explosion
- Virtual worlds, cyberspace dramas
- Consciousness and mind manipulation

More information about this series at http://www.springer.com/series/11657

Kristine Larsen

Particle Panic!

How Popular Media and Popularized Science Feed Public Fears of Particle Accelerator Experiments

 Springer

Kristine Larsen
Geological Sciences Department
Central Connecticut State University
New Britain, CT, USA

ISSN 2197-1188   ISSN 2197-1196   (electronic)
Science and Fiction
ISBN 978-3-030-12205-8   ISBN 978-3-030-12206-5   (eBook)
https://doi.org/10.1007/978-3-030-12206-5

Library of Congress Control Number: 2019931911

Cover illustration: Particle Collision and Blackhole in LHC (Large Hadron Collider) - Illustration. By general-
fmv/shutterstock.com
Photo credit: Richard Berry

This Springer imprint is published by the registered company Springer Nature Switzerland AG.
The registered company address is: Gewerbestrasse 11, 6330 Cham, Switzerland

Introduction: Anti-science Accelerator

The senile and dangerous scientist Professor Farnsworth of the animated television series *Futurama* (1999–2013) orders a do-it-yourself IVNUK SUPPERCØLLIDR from π-kea that promptly explodes when it is turned on. Preventing the explosion of a tiny particle accelerator set as a defensive system inside a room-sized holographic city taxes the talents of Nikola Tesla of the science fiction/fantasy series *Sanctuary* (2008–11) when it senses his vampire blood (it's complicated). An ultra-compact particle accelerator is also turned into a defensive weapon in the finale of the quirky science fiction series *Lexx* (1997–2002). Set to explode when it reaches the energy required to create the Higgs boson, the accelerator is launched into the heart of an Earth-threatening asteroid. The Superconducting Super Collider (SSC) never even got a chance to be completed, let alone search for the Higgs boson. In Herman Wouk's 2004 satirical novel *A Hole in Texas*, former SSC particle physicist Guy Carpenter is drawn into an international controversy when a Chinese team (led by his former lover) claims to have discovered the Higgs boson. While the U.S. government is obsessed with the possibility of being lapped in science as well as the potential military implications of the discovery, Guy's lawyer is concerned that even though scientists think a "Boson Bomb" is ridiculous, they are all "caught in a paroxysm of media imbecility" [1]. Law professor Jules Berkovits explains to Guy that the Higgs particle scares people because the "general public can't possibly understand it, and that gives it an aura of mystery and fear, the steak and potatoes of alarmist journalism" [2].

Cosmologist and science writer Sean Carroll calls particle physics "a curious activity. Thousands of people spend billions of dollars building giant machines miles across, whipping around subatomic particles at close to the speed of light and crashing them together, all to discover and study other subatomic particles" [3]. It may be curious, but particle physics has seemingly captured the imagination of science fiction authors, conspiracy theorists, and screenwriters of Hollywood blockbusters and television series. The trope of the all-powerful (and therefore dangerous) particle accelerator is so widespread in popular culture that the website *TV Tropes* has crowned it the "Phlebotinum du Jour," referring to an imaginary or impossible technological plot device [4]. In the hands of writers and directors, fictional particle accelerators can seemingly do anything (except, apparently, stay out of trouble). While it can be argued that any public interest in science (including fictional representations) is positive in the abstraction, one cannot help but notice that all the examples cited above accentuate a potential destructive power of the machines and their experiments rather than their scientific purpose—to make profound discoveries about the basic structure of our universe.

On the other side, we have the inevitable parodies, especially of the most famous particle accelerator of all, CERN's Large Hadron Collider (LHC), and the media-driven paranoia concerning its record-breaking energies. For example, two days after the July 4, 2012, announcement of the LHC's discovery of the Higgs boson, *The Daily Beast* honored the experiment as the "Meme of the Week," featuring images from around the Internet [5]. An article posted by the satirical site *The Onion* suggested that CERN scientists had become bored one year after the much-heralded Higgs discovery and were reduced to tossing random items, such as pennies, into the machine in order to entertain themselves [6]. In an obvious nose-thumbing at conspiracy websites accusing CERN of endangering the planet, the *Has the Large Hadron Collider Destroyed the World Yet?* website consists of a single word in white type on a black background: "Nope." [1]

---

[1] http://hasthelargehadroncolliderdestroyedtheworldyet.com/

Fig. 1 An aerial view of CERN with the nearly 17-mile-long underground ring of the Large Hadron Collider traced out (Maximilien Brice/CERN, CC BY-SA 3.0, via Wikimedia Commons)

Particle accelerators in popular culture may grant superpowers or promise to provide clean energy, but more often than not only succeed in destroying a city or country, the entire Earth, or even the universe itself. One might question whether such gloom and doom pronouncements by the popular media have any effect on the average citizen's views of real science and the scientists who perform these experiments. Science journalist Ian Sample reports that two-thirds of those who took part in a voluntary online BBC poll, and nearly as many in an AOL survey (both taken around the start-up time of the LHC), voiced the opinion that the machine posed too great a danger to be turned on. However, he warns, since the respondents took the time to take part in the polls, this data should not be taken as an accurate reflection of popular opinion at the time [7]. While this is certainly true, I would turn this around and remind us all that the fact that so many people took the time to voice an opinion against the LHC, based on fears that reflect much of the media hype of the time, demonstrates the power of the media to negatively affect the opinions of the general public and promulgate unhealthy and unnecessary fear.

While numerous scientists and science writers have attempted to correct factual errors in popular culture by penning "The Real Science of" books (including *Star Trek* [Lawrence Krauss], Phillip Pullman's *His Dark Materials* trio of novels [John and Mary Gribbin], and *Game of Thrones* [Helen Keen]), it is not the simple errors that have the potential to cause the most damage. At the heart of the problem is the continual creation and propagation of inaccurate (and all-too-often overtly negative) images of science and scientists, including painting scientific experiments as inherently dangerous and scientists as either simply irresponsible or actively bent on destroying the planet [8]. As this book will explore, perhaps no branch of science is more susceptible to this negative press than particle physics, especially experiments involving particle accelerators.

Of course, some in the general public have looked upon the scientific establishment with suspicion ever since the publication of Mary Shelley's *Frankenstein* in 1818, and potential misuses and ethical quandaries surrounding modern marvels such as nuclear energy, nanotechnology, and genetic engineering only fuel this public mistrust of science. A portion of the general public worries, based on what surveys demonstrate is an incomplete understanding of the basic science, if the Large Hadron Collider will create a black hole that could destroy the Earth. They question the wisdom of maintaining stockpiles of smallpox and anthrax in high security laboratories and voice unease at manipulating the genetic structures of bacteria, crops, and livestock. Each scientific advance brings with it yet another theoretical opportunity for humanity to destroy itself, whether through nuclear holocaust, pandemic, or physics experiment gone awry. Similar to popular culture's obsession with zombies, media that portray particle accelerators as death machines reflect perceived dangers that the public worries may befall humanity as a result of cutting-edge scientific research.

How should the scientific community respond to the inaccuracies and exaggerations of their research that are legion in pop culture? Since nature abhors a vacuum, perhaps scientists should simply take matters into their own hands and write engaging, positive, and factually correct fiction and nonfiction aimed at a nontechnical audience. However, this is easier said than done. For example, astrophysicists Gregory Benford and J. Craig Wheeler, and physicists David Brin and John Cramer, sadly resort to stereotypes of egotistical, unethical, and/or simply careless scientists conducting dangerous experiments in their own particle accelerator-based science fiction novels.

A common tactic is for scientists to engage in public outreach, including giving talks at libraries and schools, granting interviews to the media, and posting articles on blogs or in popular magazines. However, crafting effective and engaging science communication is complicated and, as with any other acquired skill, demands time, effort, and mentoring to fully master. In addition, communication must be tailored to its intended audience, negating the possibility of an easy one-size-fits-all template. Finally, writing for the general public is stylistically very different from the peer-reviewed journal articles most scientists have experience writing. In particular, the standard journal format of background-methodology-conclusion (tersely and tightly written with merciless and often cost-incurring word limits in mind) must be inverted and subverted in popular level works, where the emphasis is heavily placed on the results [9].

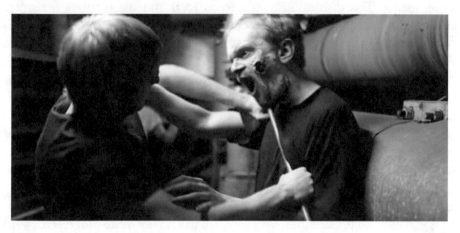

Fig. 2 A physics graduate student fights a zombie created by the Large Hadron Collider in *Decay* (2012) (H2ZZ Productions, CC-BY-NC; screen capture by author) (Public domain)

Like numerous other science professors, depictions of science in popular culture have played an important role in my teaching and outreach for years. From illustrations of science ethics in zombie media to observations of the night sky in the *Harry Potter* franchise and J.R.R. Tolkien's Middle-earth tales, Andrzej Sapkowski's pointed ecological themes in his *Witcher* series of novels to depictions of Neanderthals in the various incarnations of *Doctor Who*, the inclusion of popular culture in science education, outreach, and communication is an effective hook for attracting and holding the interest of the general public. I won't lie—a number of my colleagues have viewed my work with suspicion and occasionally open derision. I also find myself correcting misconceptions about the popular culture itself (for example, assumptions that Tolkien was anti-science [10]). But I count myself in good company; best-selling science writer Timothy Ferris wrote that his good friend Carl Sagan's success at popularizing science "stirred up a fair amount of grumbling among academics," who held the opinion that such work would only "distract scientists from rigorous research and seduce them into vulgarizing science for popular audiences"[11]. Fortunately, there is an increasing body of "rigorous research" on public receptions of science, demonstrating that Sagan has the last laugh at his critics.

It is my assertion that the scientific community's difficulty in effectively communicating with the public fuels wrong views of science and scientists, which in turn allows pseudosciences and conspiracy "theories" to flourish. This fundamental lack of understanding feeds (mainly undue) fears, which further leads the public to mistrust scientists and their big, scary, high-energy experiments. This distrust not only leads to reasonable requests for safety studies, but potentially demands by politicians and popular opinion for ultimate control over what experiments scientists are permitted to conduct. Fearing such limitations being imposed upon them by non-specialists, scientists may intentionally self-censor their communications with the public as they worry more about scaring the public than about being truthful. As I will argue, popular culture's preempting of the particle accelerator as a death machine reflects this cycle and actively takes part in furthering it. However, if we take the time to listen, popular media also suggests a way to break the chain reaction—the adoption of patient, respectful, engaging, and, well, *popular* communication with the public. When we listen to what popular culture is trying to tell us and make it an ally rather than an enemy or merely a source of aggravation to the scientific endeavor, we will finally make progress forward in replacing fear and misconception with wonder and curiosity.

This book is a study of high-energy collisions and what we can learn from their aftermath, not the literal collisions between protons or other subatomic particles, but the metaphorical collisions between fact and fiction, science and belief, public fears and professional safety studies. It is, at its core, a story about communication and what happens when we fail to communicate effectively and respectfully with each other.

In the name of transparency, I should explain at the outset what this work is *not*; it does not pretend to be an encyclopedic documentation of all appearances of particle physics and accelerators in popular culture (and my apologies if your favorite example has been left out). Through this survey we will begin to see a way forward toward effective communication with the general public concerning this fascinating and important field of human endeavor. Plot summaries will vary in length and will only include information relevant to the topic under consideration (although significant spoilers abound). My intention is to survey a wide selection of such references and demonstrate what we can learn about both public perceptions of particle physics and, perhaps more importantly, the source of unfounded fears. Along the way we will consider many different types of science "fiction," including misconceptions and erroneous claims. Our journey will sometimes proceed in a linear manner, accelerating step by step like a linac machine. More often, however, we will return to a topic, circling around for another pass like a synchrotron.

Chapters 1 and 2 lay out the general foundation of our discussion with an overview of the Standard Model of particle physics, the history of particle accelerators, and stereotypes of scientists. Chapters 3–5 focus on general fears and misconceptions about the universe, particularly as related to particle physics, including risk assessment and overall concerns raised by the start-up of large colliders at Brookhaven and CERN. Chapters 6 and 7 investigate some of the most common "end of the world" scenarios involving particle accelerators. The remainder of the work returns, like the beam of a supercollider, to the central points introduced in this introduction, reflecting on the popularization of particle physics and its potential role in overcoming some of these aforementioned fears and misconceptions.

References

1. H. Wouk, *A Hole in Texas* (Little, Brown and Co., New York, 2004), p. 173
2. H. Wouk, *A Hole in Texas* (Little, Brown and Co., New York, 2004), p. 179
3. S. Carroll, *The Particle at the Edge of the Universe* (Plume, New York, 2012), p. 7
4. Magical Particle Accelerator, TV Tropes, https://tvtropes.org/pmwiki/pmwiki.php/Main/MagicalParticleAccelerator
5. Meme of the Week: The Large Hadron Collider, The Daily Beast, https://www.thedailybeast.com/meme-of-the-week-the-large-hadron-collider
6. Bored Scientists Now Just Sticking Random Things Into Large Hadron Collider, The Onion, https://www.theonion.com/bored-scientists-now-just-sticking-random-things-into-l-1819595684
7. I. Sample, *Massive* (Basic Books, New York, 2010), p. 160
8. D.A. Kirby, Scientists on the Set: Science Consultants and the Communication of Science in Visual Fiction, Public Underst. Sci. **12**, 263 (2003)
9. Communicating to Engage, American Association for the Advancement of Science, http://www.aaas.org/comm-toolkit

10. K. Larsen, Medieval Organicism or Modern Feminist Science? Bombadil, Elves, and Mother Earth, in *Tolkien and Alterity*, ed. by C. Vaccaro, Y. Kisor (Palgrave MacMillan, Cham, 2017), pp. 95–108

11. T. Ferris, The Risks and Rewards of Popularizing Science, The Chronicle of Higher Education, https://www.chronicle.com/article/The-RisksRewards-of/77314

Contents

1

A Whirlwind Tour of Particle Physics

1.1   The Standard Model of Particles and Forces

Herman Wouk's *A Hole in Texas* (2004) begins with an author's note: "At rough guess, 99.9999% of all Americans don't know what the hell a Higgs boson is" [1]. The percentage may be comparable today, but thanks to the well-advertised discovery of the particle at the Large Hadron Collider (LHC) in 2012, a large percentage of the general public will have at least heard of it. Unfortunately, hearing about a scientific discovery doesn't necessarily make it less scary. Quite the contrary, in some cases. Various subatomic particles have made guest appearances in popular culture for decades. For example, in the 1964 episode of *The Outer Limits* "Production and Decay of Strange Particles" the introductory narration invokes the "strange world of subatomic particles", including "anti-matter composed of inside-out material, shadow-matter [neutrinos] which can penetrate 10 miles of lead shielding. Hidden deep in the heart of strange new elements are secrets beyond human understanding. New powers, new dimensions" [2]. In this chapter we will embark on a whirlwind tour of both the basics of particle physics and the enormous, complex machines used to test its predictions, including the possibility of "new dimensions". The interested reader is encouraged to consult the works referenced in this chapter if a deeper dive into the physics is desired.

Particle physics is quite simply the scientific study of the fundamental building blocks of nature, as well as their interactions via the four fundamental forces. As we were taught in school, molecules are made of atoms, and atoms in turn are composed of electrons, protons, and neutrons. Scientists wondered if these particles are truly fundamental, or made of even smaller

© Springer Nature Switzerland AG 2019
K. Larsen, *Particle Panic!*, Science and Fiction,
https://doi.org/10.1007/978-3-030-12206-5_1

particles. Like a set of Ukrainian nesting dolls, physicists have peeled back the layers of nature one level at a time, finally arriving at what is now suspected to be its most basic foundation. Much of this understanding has its genesis in the 1950s and 1960s, as physicists collided atomic nuclei and other particles at higher and higher energies, generating an entire zoo of heavier cousins to protons and neutrons. These *hadrons* (from the Greek for "thick") were suggested by physicist Murray Gell-Mann to be made of even smaller particles that he dubbed *quarks*, heralding a new classification system of particles.

One of the most counterintuitive and mysterious branches of twentieth century science, *quantum mechanics*, governs the rules of this new *Standard Model* of particles. To the non-scientist, quantum effects may appear to be more the fiction of Hollywood than the stuff of staid science. Gone is the comfortable, deterministic Newtonian view of reality as linear cause and effect, replaced by a labyrinth of potentiality. In the quantum realm the seemingly obvious distinction between a wave and a particle breaks down, and it is impossible to simultaneously know with certainty an object's motion and position (according to the *Heisenberg uncertainty principle*). Likewise, the energy bank account of the universe is fuzzy at best over very short time scales. The result is that it is possible for the universe to fudge the sacrosanct principle of conservation of energy for a brief period of time, creating energy out of what most people consider the ultimate nothingness—the vacuum of space. This borrowed energy generates a *particle–antiparticle virtual pair*, the term "virtual" applied because the couple only enjoys an ephemeral existence. In the blink of an eye, the pair annihilates each other, and in the process pays back the energy debt the universe accrued in their creation.

Yes, Virginia, antimatter really does exist, and isn't merely the product of the fertile imagination of *Star Trek* creator Gene Roddenberry or *Angels & Demons* author Dan Brown. All particles have corresponding antiparticles, with the same mass, spin, and lifetime, but opposite electric charges. Some particles, such as the photon—a particle of light—act as their own antiparticle. Similar to the process that generates virtual pairs, energy other than from the vacuum (say, from the collisions of two particles in an accelerator) can be converted into a particle–antiparticle pair, and when a particle and its antiparticle annihilate, they are converted back into energy (generally in the form of an x-ray or gamma ray). In the unimaginably high temperatures (high energies) of the early history of our universe, this yin-yang seesaw of pair creation/annihilation occurred constantly, until an as-yet-undetermined property of the laws of nature led to a tiny but important imbalance of matter over antimatter. To see antimatter today, we often need to expend energy, and a lot of

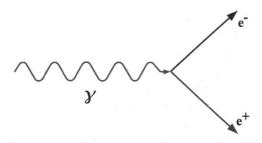

Fig. 1.1 A photon converts into an electron/positron pair (Christian Nölleke, CC-BY-SA-3.0, via Wikimedia Commons)

it.[1] As a result, despite the fact that the existence of antimatter was first theorized in 1928, the first antielectron (or *positron*) was not produced until 1932 and the first antiproton even later in 1955 (Fig. 1.1).

Another important prediction of quantum mechanics that plays a central role in particle physics is the *Pauli exclusion principle*. Simply put, each electron occupies its own quantum state, just as patrons to a theater or a baseball stadium are each assigned a unique seat. The particles that make up matter, such as electrons and quarks, obey the Pauli exclusion principle, and are collectively called *fermions*. *Bosons*, such as those associated with the fundamental forces of nature, have no such restriction. These forces are gravity, electromagnetism, the *weak nuclear force* (or simply weak force), and *strong nuclear* (or strong) *force*. The aforementioned hadrons are particles that are made of quarks and interact with each other via the strong force. The strong force itself holds the quarks together to make the hadrons. Since this force is so strong that it overcomes the electric repulsion of quarks of the same charge, it earns its name. The strong force is 137 times stronger than the electromagnetic force (which encompasses electricity and magnetism), its nearest competitor. A nod to this important ratio is found in Robert J. Sawyer's 1999 novel *Flashforward* (and its short-lived television adaptation), in the 137 second duration of the blackouts during which people see visions of the future.

Hadrons can be divided into *baryons* (from the Greek for "heavy"), which include protons and neutrons, as well as mesons, the last of which do not enter into our discussion. Since the vast majority of the mass of an atom is due to its neutrons and protons, what we consider "normal" matter is referred to as *baryonic matter*. Each baryon is composed of three quarks, held together by bosons called *gluons*. Quarks come in six varieties, called flavors, termed up,

---

[1] Or we observe high-energy events in space.

down, strange, charm, top, and bottom.[2] A proton is composed of two up quarks and a down quark, and a neutron is made from an up and two downs. The gluons are said to carry the strong force, but in what can be seen as an interesting bit of anthropomorphism, physicists also speak of bosons as "mediating" their respective forces, as if they were negotiating a contract dispute between other particles. When the universe was a few millionths of a second old it was too hot for the gluons to do their job of binding quarks together. Instead, the universe was filled with a soup of individual quarks and gluons at an unimaginable temperature of a few trillion degrees (referred to as the *quark–gluon plasma*) until it cooled sufficiently for quarks to finally remain bound together to form neutrons and protons.

The weak force is about 10,000 times weaker than the strong force, and describes how quarks change flavor as well as interactions involving electrons and their cousins, collectively called *leptons* (from the Greek for "light"). Just as there are six flavors of quarks, there are six different types of leptons—the electron, muon, and tau as well as three corresponding neutrinos. Unlike hadrons, leptons are truly fundamental particles (in other words, they are not made of smaller particles). Weak force interactions involve both quarks and leptons, and because these interactions can sometimes involve a change in electric charge, three different bosons mediate the weak force—W^+, W^-, and the neutral Z. This last boson has similarities to a photon, the particle that mediates the electromagnetic force, although the Z has a nonzero mass (Fig. 1.2).

The family resemblance between the Z and the photon suggested to physicists that in high-energy experiments (at a temperature of around 10^{15} or a million billion degrees) these similarities would become even more overwhelming and the two forces would act as one, dubbed the *electroweak force*. If the electromagnetic and weak forces unify at high energies, could the same be possible with the strong force? Although there is no one unique unification scheme, a class of solutions called *grand unified theories* (or GUTs) predict exactly this. Part of the problem is the unimaginable energy necessary to directly test the predictions of GUTs. In order to experimentally generate the necessary temperature (around 10^{28} degrees, a 1 followed by 28 zeroes), a particle accelerator over seven light years long (more than one and a half times the distance to the nearest star system) would be required, making such an experiment unlikely, to say the least [3]. Instead, particle physicists work with cosmologists, scientists who study the history and structure of the universe, in using observations of the universe itself to test these predictions.

---

[2] In a touch of whimsy, the t and b quarks are sometimes referred to as truth and beauty.

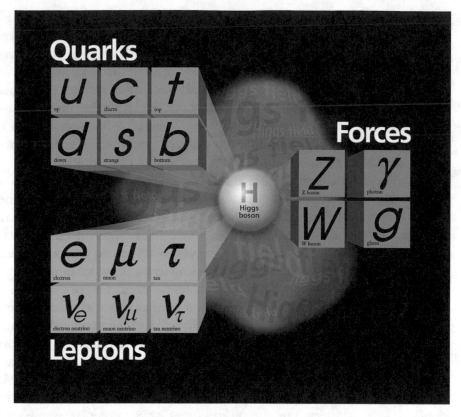

Fig. 1.2 The fundamental particles of the Standard Model (Public domain)

If three of the four fundamental forces can be unified in the high temperatures and energies of the early universe, why not all four? Physicists expect
that they are, just don't ask them to explain exactly *how*. The problem is that
the three forces combined in GUTs all obey the rules of quantum mechanics.
The fourth fundamental force, gravity, is currently described by Einstein's
general theory of relativity, whose basic formalism is very distinct from that of
quantum mechanics. For example, in general relativity the gravitational force
is described by the warping of the four dimensional "fabric" of the universe,
called *space-time*, that encompasses our three dimensions of space and one of
time. In a quantum model of gravity the force would be said to be carried by
a boson called a graviton (similar to the other three forces). A successful marriage between quantum mechanics and general relativity would result in a
completely unified *theory of everything* (TOE) and would explain the earliest
moments of our universe's history.

Putting aside the problem of gravity, there is one final important piece to the Standard Model, the infamous *Higgs boson*. In John Cramer's 1997 novel *Einstein's Bridge*, egotistical experimental physicist Jake Wang brags to visiting writer Alice Lang that she has arrived at the "brink of the most momentous discovery of our new century. The Higgs particle… this magical particle is about to reveal itself to us" [4]. What is it about the Higgs that made its discovery the holy grail of particle physics and resulted in the awarding of a Nobel Prize to Peter Higgs and François Englert a mere year after its discovery? To explain this we need to introduce the concept of a quantum field.

A *field* is a physical quantity that can be defined over a region of space-time. The best-known examples are electric fields and magnetic fields (under the generic umbrella of electromagnetic fields), and gravitational fields. Fields are invisible, but they affect matter and therefore can be measured and studied. For example, a charged particle such as a proton will move in a predictable way in the presence of a magnetic field (an important principle in the design of particle accelerators). In quantum mechanics the excitations of a given field (like ripples in a pond) are associated with a particular particle with certain properties. An example is the electromagnetic field and the associated photon.

Physicists expect that the four forces of nature were unified at the high temperatures of the very early universe, and only took on their unique identities as the universe expanded and cooled. This initial state of unification is called a *symmetry*, and the process by which the forces take on their distinctive identities is termed *spontaneous symmetry breaking*. An everyday example of symmetry breaking can be found at a round dinner table. Initially there is a perfect symmetry in the positions of the plates and glasses around the table. However, all it takes is for one person to choose whether to use the wine glass on their left or right to break the symmetry for the entire table. If the first person selects the glass to the left, so must everyone else at the table. Symmetry preserving states of quantum fields are unstable, meaning that a state of broken symmetry is preferred by nature because it requires less energy. Think of it this way—the guests at the dinner table feel less stressed once someone has broken the symmetry and decided whether everyone will use the wine glass to their left or right (Fig. 1.3).

The Higgs boson enters the fray when discussing a symmetry involving the weak force. In a state of perfect symmetry, quarks, electrons, muons, taus, and the W and Z bosons would all be massless, like the photon.[3] But in our current universe, they are not only massive, but each has a distinct mass (except

---

[3] The source of the small masses of the neutrinos may be the same as for these particles, or it may not. Stay tuned as physicists figure this out.

Fig. 1.3 The symmetry of this dinner table will be broken when the first guest decides which wine glass to use (Public domain)

for the photon). Similar solutions to this problem were independently proposed in 1964 by the team of Robert Brout and François Englert as well as Peter Higgs, now collectively called the *Brout-Englert-Higgs mechanism*. The idea is that there exists a special field (alternately called the BEH or Higgs field) that had a value of zero in the extremely hot early universe. As the universe cooled to a specific temperature, the BEH field took on a nonzero value, and through its interactions with electrons, quarks, and the rest, breaks the symmetry in mass and assigns each one its individual mass. Higgs went further in proposing the existence of an associated particle, a boson that now bears his name [5]. Physicists began to search for evidence of the Higgs boson in the mid-1970s using accelerators with higher and higher energies, necessitated by the presumed large mass of the Higgs (now known to be about 126 GeV,[4] or about 134 times heavier than a proton). In July 2012 CERN announced that the LHC had confirmed the existence of the Higgs boson, and therefore the validity of the BEH mechanism. Englert and Higgs shared the 2013 Nobel Prize in Physics (Brout being deceased by that time).

---

[4] These energy units (based on the electron volt) will be explained in more detail in Chapter 3. An MeV is a million eV (electron volts), a GeV is a billion eV, and a TeV is a trillion eV. While a trillion anything sounds like it should be very large, the energy of motion (kinetic energy) of a mosquito buzzing around your head is about 1 TeV.

With the discovery of the Higgs boson, the existence of all the predicted particles of the Standard Model were confirmed. However, there still remain a number of unresolved questions. One involves the nonzero value of the Higgs field itself (the so-called *vacuum expectation value*, or VEV). At 246 GeV it is relatively small, especially compared with the so-called Planck scale (about 10^{19} GeV), the energy scale of the eventual TOE (the energy of the early universe at the point where gravity and quantum mechanics must be combined to provide an accurate description of its properties) [6]. This is known as the *hierarchy problem*. The nonzero value of the Higgs field VEV sets the lowest possible energy state of the universe, called the energy of the vacuum. But since the value is not zero, it is not considered a true vacuum, but rather a false vacuum. Is it possible that the universe will eventually decide to transition to a true vacuum state? What would that mean for humanity? And why do some people worry that the LHC might trigger this hypothetical transition? We will consider these questions in due time.

Both the hierarchy problem and the lack of a TOE are clues that our understanding of the universe and its early moments is incomplete. One proposition for achieving the unification of the forces is through connecting the particles that carry the forces (bosons) with those that make up matter (fermions). In *supersymmetry* (called SUSY for short), every boson is associated with a fermion, and vice versa. These hypothetical associated particles are called *supersymmetric partners*. For example, the electron, a fermion, would have a supersymmetric boson partner called the selectron. Versions of SUSY that include gravity are called *supergravity* (or SUGRA), where gravity is thought to be mediated by a boson called the graviton. Its SUSY partner is a hypothetical fermion called the gravitino. The photino, the hypothetical supersymmetric partner to the photon, plays a central (and nefarious) role in Stephen Baxter's 1994 novel *Ring*, where it is responsible for killing the Sun. Because the supersymmetric partners are hundreds or thousands of times heavier than the particles of the Standard Model they would require collisions of tremendous energy to produce in the lab, and none have yet been discovered. However, if and when they are found, their specific masses might explain the small value of the Higgs field VEV [6]. They may also provide an explanation for another enigmatic resident of our universe, *dark matter* (discussed in Chapter 6). Physicists are wisely not putting all their eggs in one basket, and are actively exploring other extensions to the Standard Model, some of which include the existence of dimensions beyond the three physical dimensions and one temporal dimension of our everyday lives.

1.2   M-Theory and Multiple Dimensions

In Franklin Clermont's novel *The Voices at CERN* (2014) and its sequel, *The Terror at CERN* (2015), the LHC demonstrates the existence of an eighth dimension where the consciousness of all humans, intelligent extraterrestrials, and even dogs, exist after death as Energy Beings (EBS).[5] While this is clearly science fiction, the potential existence of additional dimensions is very much rooted in science. As previously noted, our current understanding (thanks to Einstein) is that space and time are interwoven into a four-dimensional fabric termed space-time. There is no known reason why additional spatial dimensions beyond our typical three (length, width, height) could not exist, although their properties would be limited by the laws of physics we observe today. In 1921, mathematician Theodor Kaluza invoked an extra dimension of space in an attempt to unify gravity and electromagnetism, and in 1926 mathematician Oskar Klein demonstrated that Kaluza's extra dimension would be rolled up, or compactified, into a circle the size of the Planck length (10^{-35} m) and hence unobservable. While this unification scheme did not succeed, it opened the door for further investigations into the physics of higher dimensions.

In the late 1960s through 1970s physicists developed models with tiny extra dimensions that became known as different flavors of string theory. In these models each unique frequency of vibration of a tiny Planck length-sized object termed a string was hypothesized to be what we observe as a different elementary particle. String models that include supersymmetry are called superstrings. The fictional physicists of Stephen Johnson's 2012 e-novel *God's Spark* propose a radical new revision of superstrings in order to explain the current lack of evidence for their existence. In the real world, M-theory is an 11-dimensional extension of superstrings that combines it with supergravity. M-theory predicts that besides point-like (zero-dimensional) particles and one-dimensional strings, there could be higher dimensional objects generically called branes. In contrast with string theory, these extra dimensions do not necessarily have to be tiny; they only need to have properties that do not defy the observed laws of nature. Braneworld models usually invoke one additional spatial dimension called the bulk, and with the right assumptions imposed, the previously mentioned hierarchy problem can be solved. If the extra dimension[6] exists, gravity would spread from our observable universe

---

[5] It is never explained what exists in the fifth, sixth, or seventh dimension (besides a 1970s musical group, of course).

[6] It is possible that there exists more than one extra spatial dimensions, but adding more dimensions further restricts their possible properties.

(the brane) into the higher dimensional bulk, and if the Planck energy in the bulk is diluted enough to be of roughly the same size as the Higgs VEV, then there is no disparity in their energies and thus no hierarchy problem. Note that there are a lot of "ifs" here, but it is an interesting possibility.

Although we would not observe this extra dimension directly, its existence could possibly be confirmed through particle accelerator experiments at energies of about 1000 GeV or 1 TeV. The LHC, operating at energies of about 13 TeV, is searching for such telltale events [7]. For example, there may exist versions of the W and Z bosons with 100 times their standard masses, called Kaluza-Klein states. Of specific interest to this discussion is the fact that if it exists, this extra dimension might make it possible for the LHC to create microscopic black holes, a potential that has created excitement and fear, depending on the audience. We will return to this in depth in Chapters 5 and 7. For now, we will turn our attention to the machines that make such experiments possible—particle accelerators—and the basic principles behind their operation.

1.3   An Accelerated History of Accelerators

Atom-smasher, a mutated human (or meta-human) on the TV series *The Flash* (2014–),[7] earns his nickname because he absorbs the radiation caused by the power of the atom to increase his size and strength and has a tendency to smash things. "Atom-smasher" is also a nickname for a particle accelerator, and this meta-human's powers were caused by a particle accelerator accident. Technically speaking, a particle accelerator doesn't smash atoms, but individual charged particles (such as electrons or protons) or the dense nuclei of atoms, often called ions. This begs the question, why would physicists intentionally cause such collisions in the first place, and what tools do they use in these experiments?

Since particle physics deals with the smallest constituents of matter, a particle accelerator is analogous to a microscope in its ability to magnify the properties of these tiny particles and bring them into focus. This is done by first accelerating the particles to high speeds, as close to the speed of light as possible, directing them to collide, and then sitting back and watching what happens. Despite what you might expect, physicists are not smashing the

---

[7] Only Seasons 1 and 2 are referenced in this book, in part because depictions of science in this series deserve their own book-length treatment, but mainly because the plot becomes exponentially more convoluted beginning in Season 3.

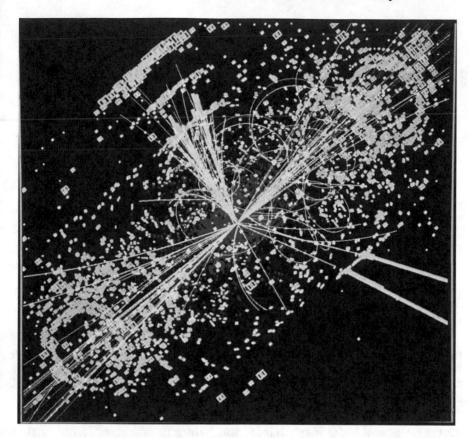

Fig. 1.4 A simulation of the complex results of two colliding protons creating a Higgs particle at the LHC (Lucas Taylor/CERN, CC BY-SA 3.0, via Wikimedia Commons)

particles together to figure out what they are made of, because we already have that information. Instead, the goal is to create other particles that we normally don't see hanging around the laboratory on their own. In many cases these particles quickly decay, so in reality we are not looking for the particle in question, but rather the decay products that confirm that we were successful in briefly generating the particle we desired. This is the case with the Higgs boson, for example. Sean Carroll likens the process to "smashing together two Timex watches and hoping that the pieces assemble themselves into a Rolex" [8] (Fig. 1.4).

Only electrically charged particles like electrons and nuclei are employed, because we use an electric field to give them their kick and accelerate them to higher and higher velocities. If the particles travel once down a straight pipe, we have a *linear accelerator or linac*. By applying a magnetic field, the path of a charged particle can be changed, for example into a circle. As a particle

Fig. 1.5 A beam of fast-moving electrons is curved into a circular path by a magnetic field. Collisions with air molecules cause the telltale fluorescent glow. A particle accelerator beam travels in a near vacuum to avoid colliding with air (Marcin Białek, CC BY-SA 4.0, via Wikimedia Commons)

moves faster and faster it gets harder and harder to accelerate it any further. This is because its effective mass gets heavier and heavier, as predicted by Einstein's special theory of relativity. For this reason we need very intense electric and magnetic fields to achieve high collision energies. These are the fundamental principles of particle accelerator physics.

The first particle accelerator was developed between 1928–1932 in the famous Cavendish Laboratory at Cambridge University by John D. Cockcroft and Ernest T.S. Walton under the supervision of famed nuclear physicist and Nobel laureate in Chemistry Lord Ernest Rutherford. A 2 meter long glass tube was used to accelerate protons downward by applying different voltages at the top, middle, and bottom. A proton would obtain a final kinetic energy of up to 0.75 MeV in this machine, and when one struck a stationary lithium nucleus at the bottom of the accelerator the result was two helium nuclei [9]. Cockcroft and Walton were awarded the 1951 Nobel Prize in Physics for their work. Unfortunately, bothersome electric sparking occurs at energies of about 1 MeV, which limits this design [10].

At about the same time, the *cyclotron*, the first design to use a magnetic field to steer the charged particles, was developed by Ernest O. Lawrence at the University of California-Berkeley. Instead of a fixed change in voltage as in the Cockcroft–Walton design, an alternating electric field, called a *radiofrequency* or RF field, was used. The cycle of this oscillating electric field is tuned to give the charged particles a series of equally timed kicks that accelerate them to higher and higher velocities. Because the magnetic field in the device is constant, the particles travel in larger and larger circular orbits as they race faster and faster, spiraling outward until they reach an exit at the edge of the disk-shaped machine [11]. Lawrence's first cyclotron was only 13 cm across and accelerated protons to a meager 0.08 MeV. Although much larger cyclotrons were built, two issues ultimately limited their size. First, the physical size of the disk-shaped magnets became a problem, and, more importantly, the increase in the effective mass of the particles at high speeds meant that the particles became sluggish and didn't reach the proper point in the machine when the electric "kick" was applied. The result is that a cyclotron cannot propel protons to energies over about 20 MeV (Fig. 1.5).

The distinctive disk shape of a cyclotron makes it obvious when it appears in popular culture. *The Boat* (*El Barco*; 2011-3) is a Spanish science fiction tv drama that has surprisingly escaped the eyes of both English-speaking audiences and media critics. Similar in structure and style to *Lost* (without the spiritual overtones), the 44 episodes follow the adventures of the teaching ship *The Polar Star*, its crew (including scientist Dr. Julia Wilson), and forty college age students. All find themselves part of the top-secret Alexandria Project meant to repopulate the Earth after a particle accelerator accident destroys all land (with the exception of a few isolated islands).

An old black and white reel-to-reel movie is found by the students documenting the fictional 1940 Barkley Symposium of scientists that had been secretly held aboard the ship. The attendees include one "T.S. Walton", said to be the creator of the first accelerator (an obvious nod to Ernest T.S. Walton). Walton hides an accelerator schematic on the ship in order to prevent it from falling into the wrong hands and being used as a weapon. In the movie soldiers are seen boarding the ship, torturing the Captain, and killing members of the crew, presumably for said schematic. The prop department for the series made the interesting choice of having the diagram be a sketch from Ernest Lawrence's 1934 patent for the cyclotron (Fig. 1.6) [12].

In popular culture, when particle accelerators aren't busy causing mayhem and destruction they are often touted as sources of clean, limitless energy (usually right before said mayhem and destruction occur). For example, *The Flash*'s scientist Harrison Wells promises his accelerator will "change our

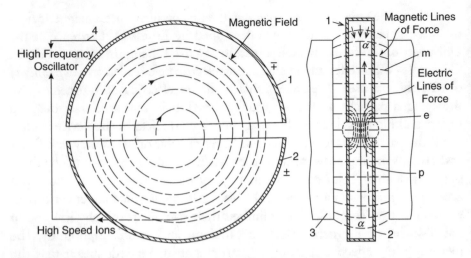

Fig. 1.6 Ernest O. Lawrence's patent design for the cyclotron (Public domain)

understanding of physics. Will bring about advancements in power, advancements in medicine" [13]. Instead, it kills over a dozen people and results in the mutation of thousands of others. Similarly, in the 2010 graphic novel *Transformers Timelines: Generation 2: Redux* the Decepticons use the LHC to refine the rare terrestrial fuel source forestonite. A battle for control of the accelerator leads to the wide release of the substance, granting the Transformers new superpowers. Dr. Thomas Abernathy, the mad scientist of director Gilbert Shilton's *The Void* (2001), promises his shareholders and the members of the Atomic Energy Commission that his accelerator will provide "Boundless supplies of clean, safe energy…. That's enough for our hospitals, our schools, our homes" [14]. They fire up the accelerator and a self-sustaining black hole results, sucking in several technicians before imploding the entire building. Director Julius Onah's 2018 direct to Netflix film *The Cloverfield Paradox* takes place in the year 2028, with the Earth in the throes of a crushing energy crisis. In response, a revolutionary particle accelerator called the Shepard is built aboard an experimental space station in Earth orbit. Its goal is to create a sustained beam that will somehow solve the energy crisis. Instead, it breaks down the walls between parallel realities, allowing monsters to invade Earth(s) and resulting in unimaginable death and destruction. In reality, particle accelerators are almost the exact opposite, as they require energy in order to accelerate particles to relativistic speeds, rather than provide energy back to society. Nevertheless, particle accelerators are often touted as sources of energy (for good or evil) in many of their sci fi incarnations.

We have already seen fictional worries about the potential use of an accelerator as a weapon in the tv series *The Boat*, but the series also plays with the idea that a particle accelerator could be an energy source as well. Aboard *The Polar Star* the students find the original captain's logbook and read about an enigmatic passenger from Liverpool given passage in 1941 with his equally mysterious top secret briefcase. After the mystery man's death during the ship's outbreak of the Spanish Flu, Captain Zuniga commits to paper the man's confidential explanation of his secret project for the Allies that is motivated by the potential dangers of atomic weapons. The Captain is frightened by the man's talk of "geological cataclysms" and muses "I don't know if I'm talking to a genius or madman" [15]. In the present day it is revealed that Zuniga's son, now an old man, had broken into the briefcase at the time and hidden aboard the ship papers about a top secret project codenamed Alexandrie. In the event of the Nazis assembling successful atomic weapons, the plan was to send out seven self-sufficient ships with handpicked crews in order to repopulate the world after the inevitable nuclear holocaust. One of these hidden papers contains a schematic of the fifth ship, the French *Etoile du Nord*, showing that it was equipped with a betatron accelerator for the (fictional) purpose of generating power.

The first successful *betatron* was built by Donald Kerst of the University of Illinois in 1940. The design uses a varying rather than a static magnetic field to curve the path of the particles. This is done by using an electromagnet (a magnetic field generated by an electric current) rather than the disk-shaped permanent magnet of the cyclotron. Unlike in a cyclotron, electrons (also called beta particles) are specifically used, and remain in a constant orbit thanks to the changing magnetic field. The electrons therefore move within a circular vacuum tube rather than a disk. Kerst's original machine was 15 cm in diameter and accelerated electrons to 2.3 MeV. Larger instruments were subsequently made, reaching energies of 300 MeV. When the high energy electrons are directed onto a metal target, the result is copious x-rays (for example for radiation treatment for cancer), which is the betatron's greatest practical use [16].

The development of the *synchrotron* was a major technological advancement in particle accelerators. The RF electric field is tuned in such a way that it keeps in step with accelerating bunches of particles (although we still call it a "beam"). Each time a bunch reaches an RF source (called a cavity) it receives a synchronized kick of energy; the magnetic field in the bending electromagnets is also increased to keep in step with the increasing velocity of the particles, keeping the circular path of the particles constant. This allows the apparatus to be a simple circular tube like the betatron, rather than the disk-shaped cyclotron. Other electromagnets called focusing magnets keep the particle bunches tightly clumped in order to increase the likelihood that collisions will occur with the target. Much higher energies can be reached in a synchrotron. For example, as early as 1946 a synchrotron accelerated deuterons (a proton and neutron bonded together) to 195 MeV [17].

The energy of accelerated particles in a synchrotron is limited by two properties, the strength of the bending magnets and the loss of energy from what is called *synchrotron radiation*. The sharper the curve in the vacuum tube (i.e. the smaller the circumference of the particle's path), the stronger the magnetic field required to keep the particles in the proper orbit. Using a large circumference ring helps with this issue, and also aids in minimizing a second problem. Charged particles accelerating in a circle emit light; in synchrotrons this is usually in the form of ultraviolet and x-rays, both of which are not exactly healthy to human tissues. The dangerous radiation can be shielded, for example, by putting the accelerator underground as in the case of the LHC, but in general a more gentle curve (a longer path) minimizes the radiation as well.

Radiation risks are apparently both serious and inconsequential to genius industrialist Tony Stark of Stark Enterprises, aka Iron Man. In director Jon Favreau's *Iron Man 2* (2010) Stark builds an accelerator in his personal workshop in order to create a new element for his Iron Man suit's arc reactor, one that won't produce harmful radiation. Apparently oblivious to the dangers of any radiation generated by his tabletop accelerator, he not only remains in the same room with the working accelerator but steers the path of the beam using a large wrench (rather than the massive magnets used in real-world accelerators) [18]. It is true that particle accelerators have long been used to create new elements. For example, plutonium (a fuel in atomic weapons) was first generated in the Berkeley cyclotron by bombarding uranium with deuterons [19]. Other heavy elements were discovered at the Joint Institute for Nuclear Research (JINR) in Russia. In "Production and Decay of Strange Particles", a 1964 episode of *The Outer Limits*, Dr. Marshall's experiment in creating heavy elements in his cyclotron by bombarding nobelium with high energy particles from outer space goes horribly awry, resulting in the deaths of all his assistants and colleagues, and the near destruction of the entire planet at the hand of hostile interdimensional creatures made from radiation.

But a particle accelerator can also save your life, if you happen to be chased by an advanced T-X model terminator. In the 2003 Jonathan Mostow film *Terminator 3: Rise of the Machines*, John Connor and his future wife run through a small accelerator facility on a military base trying to escape their metallic nemesis. Connor powers up the electromagnets as the T-X chases the couple around the accelerator ring. When the magnets are fully powered everything metallic, including John's machine gun and the T-X, are yanked against the magnets, stopping her until she incapacitates the magnetics by cutting through their all-important coolant system. The T-X was apparently the temporary victim of a synchrotron type accelerator using superconducting magnets that need to be cooled to nearly absolute zero in order to operate efficiently. What happens in real life when you release the coolant required by such magnets is closer to the stuff of a disaster movie, as we shall explore in Chapter 5.

Heavier particles, such as protons, lose less energy through synchrotron radiation than do lighter electrons. Does this mean that physicists prefer proton accelerators? The answer is, it's complicated. Recall that protons are made of quarks, which are held together by gluons. When protons collide, it's not like two hard billiard balls colliding (as in the case of electrons), but instead closer to two teddy bears. You might get lucky and two hard eyes might collide, but the stuffing will almost always play some role. The good news is that each collision will be a bit different. If you don't exactly know what you are looking for (as in the case of the search for the Higgs boson, whose mass was not known ahead of time), if there is some interesting phenomenon to be found you will hopefully find it if you collide enough teddy bears together [20]. Electrons are more suited to the case where you know what you are looking for, and can arrange for the collisions to occur at the predicted energy (for example, to generate a particle of predictable mass).

Whichever particles we use, we can increase the energy if we collide two particle beams moving in opposite directions at the same energies rather than crash a single beam into a stationary target. For this reason most modern large particle accelerators are "colliders". An example is the Tevatron, a synchrotron particle accelerator at Fermilab in Batavia, Illinois operational between 1983 and 2011. Until the LHC came online in 2008 it held the crown as the world's highest energy collider. Protons and antiprotons circled its 3.9 mile circumference ring in opposite directions at nearly the speed of light, producing collision energies of up to 2 TeV. While the LHC in Europe ultimately bested these energies, the Tevatron potentially had more local competition, at least on paper, the Superconducting Super Collider (SSC), planned for Waxahachie, Texas.

1.4   Supercolliders: RIP SSC and Hello LHC

John Cramer's novel *Einstein's Bridge* takes place in an alternate timeline in which the SSC became a reality. In the novel protons accelerate around a 54 mile circumference tunnel, collide with a whopping 40 TeV, and successfully create Higgs particles. Regrettably, the high-energy experiment alerts the hostile extraterrestrial species the Hive of humanity's existence. Despite the efforts of the benevolent ET species the Makers to help humanity defend the Earth, all that can be done is to send two of the accelerator scientists back in time (conveniently after teaching them how to genetically engineer their appearance and aging). Their goal is to prevent the SSC from being created in the first place and thereby prevent the Hive's attack.

The true story of the demise of the SSC does not involve hostile extraterrestrials, but rather hostile politicians and cost overruns. The project was initially proposed in 1982 in order to allow America to regain the particle physics crown after the discovery of the W and Z bosons at CERN. With a proposed cost of 2.9 to 3.2 billion dollars, and promising not only myriad jobs for scientists but support staff and construction workers (not to mention much-needed spin-off business for the local community), several sites across the country vied for the honor of hosting the project. The eventual Texas site was steeped in politics (as Texans George H.W. Bush and Jim Wright were then Vice-President and Speaker of the House, respectively), the budget mushroomed to an estimated 12 billion dollars as the project progressed, and the proposed International Space Station competed for Congress's attention and dollars. In the end Congress decided to pull the plug in 1993 after 2 billion dollars had been spent and 14 miles of tunnel excavated [21]. With the SSC off the table, CERN's ruling Council voted in December 1994 to build its own "super" collider, the LHC.

The name "Large Hadron Collider" is indelibly ingrained in geek culture. For example, in the 2010 *The Big Bang Theory* episode "The Large Hadron Collision" experimental physicist Leonard Hofstadter plans to attend a conference at CERN on Valentine's Day, and his buddy particle physicist Sheldon Cooper is crushed that Leonard is planning to bring his waitress girlfriend along instead of his friend. In Sheldon's plaintive words, "I've been dreaming about going to the Large Hadron Collider since I was nine years old" [22]. The actual title of the LHC's home institution, Conseil Européen pour la Recherche Nucléaire (European Council for Nuclear Research), rolls off the tongue less well than its snappy acronym, CERN. Established in 1954, it currently has 22 member nations and plays the roles of employer or host to thousands of scientists and engineers at its facility on the France-Switzerland border near Geneva. It was here that the World Wide Web was developed in 1989, originally as a means of expediting communication between scientists; the world's first website, http://info.cern.ch, is currently being restored by CERN to preserve its history.

The LHC was built using the existing 16.6 mile circumference underground tunnel constructed for an earlier project, the Large Electron-Positron Collider or LEP (1989–2000). LEP accelerated and collided electrons and anti-electrons (positrons), producing collision energies over 200 GeV and leading to the discovery of the W and Z bosons [23]. Most of the LHC's collisions involve protons against protons, with collision energies now up to 13 TeV, the highest ever achieved in an accelerator. Much heavier lead nuclei can also be collided with each other and protons, resulting in different particle products. In particular, lead–lead collisions mimic the high temperature, high-density conditions found in exploding stars (supernovae) and the quark–gluon plasma (Fig. 1.7).

Fig. 1.7 The ALICE particle detector at the Large Hadron Collider (John-vogel, CC BY-SA 3.0, via Wikimedia Commons)

The extreme energies reached at the LHC are only possible thanks to its 9600 powerful high-tech magnets. To achieve the necessary magnetic fields while using as little electricity as possible, the magnets' superconducting material is cooled to the extremely low temperature of 1.9 K ($-456°$ F). This is accomplished using massive liquid helium cryostat containers [24]. The results of the particle collisions are studied by a host of detectors. ATLAS and the Compact Muon Solenoid (CMS) are the general-purpose detectors that independently discovered the Higgs boson. ALICE (A Large Ion Collider Experiment) analyzes the properties of the quark–gluon plasma created in lead collisions, while Large Hadron Collider beauty (LHCb) investigates the interactions of particles containing the b quark in order to help explain why matter won out over antimatter in the early universe [25]. As in the case of any large particle accelerator, every few years the LHC is taken offline for maintenance and upgrades. For example, after its February 2013 through May 2015 shutdown it finally achieved its maximum collision energy of 13 TeV. The current shutdown is planned for December 2018–Feb 2021 [26]. There was also an infamous unplanned 14 months shutdown 9 days after its initial startup in September 2008 due to an "incident" with the magnets. Such "oops" moments feed into public scrutiny of the LHC's safety (as we shall explore in Chapter 5). Having completed this circuit through the basic physics, our next stop along the metaphorical collider tunnel brings us to depictions in popular culture of the individuals who study these collisions, the scientists themselves, including issues of gender.

References

1. H. Wouk, *A Hole in Texas* (Little, Brown and Co, New York, 2004), p. 3
2. L. Stevens (script), Production and decay of strange particles, The Outer Limits, season 1 (1964)
3. A. Guth, *The Inflationary Universe* (Helix Books, Reading, 1997), p. 31
4. J. Cramer, *Einstein's Bridge* (Avon Books, New York, 1997), p. 86
5. V. Jamieson, What to call the particle formerly known as Higgs. New Scientist, https://www.newscientist.com/article/dn21604-what-to-call-the-particle-formerly-known-as-higgs/
6. K. Garrett, G. Duda, Dark matter: a primer, https://ned.ipac.caltech.edu/level5/March10/Garrett/paper.pdf
7. G. Landsberg, Collider searches for extra dimensions, SLAC Summer Institute on Particle Physics (SSI04), Aug. 2–13, 2004, http://www.slac.stanford.edu/econf/C040802/papers/MOT006.PDF
8. S. Carroll, *The Particle at the Edge of the Universe* (Plume, New York, 2012), p. 56

9. Y. Ne'eman, Y. Kirsh, *The Particle Hunters*, 2nd edn. (Cambridge University Press, Cambridge, 1996), p. 89

10. P.J. Bryant, K. Johnsen, *The Principles of Circular Accelerators and Storage Rings* (Cambridge University Press, Cambridge, 1993), p. 3

11. W.J. Kernan, *Accelerators* (U.S. Atomic Energy Commission, Oak Ridge, 1968), pp. 12–13

12. M. Cistaré, J. Naya (script), The Mermaid, The Boat, season 2 (2011)

13. A. Kreisberg, G. Johns (script), Pilot, The Flash, season 1 (2014)

14. G. Shilton (dir.), The Void, Lions Gate Entertainment (2001)

15. H. Ortega, J. Reguilón (script), The Man from Liverpool, The Boat, season 1 (2011)

16. A. Sessler, E. Wilson, *Engines of Discovery: A Century of Particle Accelerators*, Rev. edn. (World Scientific, Singapore, 2014), p. 47

17. J. Allday, *Quarks, Leptons, and the Big Bang*, 2nd edn. (Institute of Physics, Bristol, 2002), pp. 184–185

18. E. McCarthy, Can You Build a Particle Accelerator in Your Home? Iron Man 2 Fact Check. Popular Mechanics, https://www.popularmechanics.com/culture/movies/a12418/iron-man-2-particle-accelerator/

19. W.J. Kernan, *Accelerators* (U.S. Atomic Energy Commission, Oak Ridge, 1968), p. 18

20. L. Evans, Particle Accelerators at CERN: from the early days to the LHC and beyond. Technol. Forecast. Soc. Change **112**, 6 (2016)

21. A. Sessler, E. Wilson, *Engines of Discovery: A Century of Particle Accelerators,* Rev. edn. (World Scientific, Singapore, 2014), p. 92

22. L. Aronsohn, R. Rosenstock, M. Ferrari (script), The Large Hadron Collision, The Big Bang Theory, season 3 (2010)

23. L. Lederman, C. Hill, *Beyond the God Particle* (Prometheus Books, Amherst, 2013), p. 185

24. L. Lederman, C. Hill, *Beyond the God Particle* (Prometheus Books, Amherst, 2013), p. 27

25. Experiments, CERN, https://home.cern/about/experiments

26. Longer Term LHC Schedule, CERN, https://lhc-commissioning.web.cern.ch/lhc-commissioning/schedule/LHC-long-term.htm

2

Angels or Demons? Stereotypes of Scientists

2.1 The Mad Versus the Noble Scientist

Sociologist Eva Flicker argues that all forms of art contribute "to the mediation of science. The images drawn of science and scientific work are deeply embedded in the culture" [1]. Arguably the most common of these images is the cartoonish mad scientist, hell bent on either proving a pet hypothesis correct or world domination (or both), no matter the cost. Legendary science popularizer Carl Sagan wrote that this image haunts our "world—down to the white-coated loonies of Saturday morning children's TV and the plethora of Faustian bargains in popular culture, from the eponymous Dr. Faustus himself to *Dr. Frankenstein*, *Dr. Strangelove*, and *Jurassic Park*" [2].

Such stereotypical depictions appear quite often in particle physics based science fiction, with even real world scientists themselves embracing the trope. In his commentary track to the film treatment of his novel *The Krone Experiment*, physicist J. Craig Wheeler pulls no punches when describing the titular character he portrays as a "brain dead evil scientist" who intentionally creates black holes that he cannot control [3]. The difficult but brilliant Jake Wang, whose accelerator detectors have "produced excellent physics, and all of them have left a trail of broken minds and bodies in their wake" [4] in John Cramer's novel *Einstein's Bridge* sounds suspiciously similar to the picture painted of Nobel Laureate Carlo Rubbia in Gary Taubes' 1986 exposé *Nobel Dreams: Power, Deceit, and the Ultimate Experiment*. In Cramer's tale disaster novel author Alice Lang worms her way into the SSC under the cover of writing a story about the facility. On a tour of the facility she witnesses one of Jake's legendary hissy fits and wonders if it was meant to "deceive her into believing that mad

© Springer Nature Switzerland AG 2019
K. Larsen, *Particle Panic!*, Science and Fiction,
https://doi.org/10.1007/978-3-030-12206-5_2

scientists could be found outside late-show sci-fi flicks" [5]. Physicist George Griffin explains to her that particle physics "attracts some of the best minds in physics, as well as some of the strongest personalities" [6]. This may be true, but where is the boundary between strong-willed and mad?

Not surprisingly, the iconic mad scientist looms large in a range of stereotypes of scientists Roslynn Haynes found in an analysis of Western literature. The other six are

1. The "evil alchemist" who works in secret labs on illegal experiments;
2. The "noble scientist", the hero or savior of society;
3. The "foolish scientist", aka the absent-minded genius;
4. The "inhuman researcher" who has sacrificed relationships, emotions, and all vestiges of humanity in the name of science;
5. The "scientist as adventurer", such as the Doctor of *Doctor Who* or Indiana Jones; and
6. The "helpless scientist" who has no malicious intent, but whose experiment simply gets out of control and threatens the world [7].

Note that only numbers 2 and 5 on this list can be read as positive depictions. The mad scientist and the evil alchemist are closely linked, and one could argue that they should be considered together. Doing so aligns well with a similar sixfold classification of scientists in film and television found in Kevin Grazier and Stephen Cass's *Hollyweird Science*. These are the mad scientist, socially awkward nerd, sidekick, helpless victim, hero, and "Conflicted Prantagonist … a mash-up of protagonist and antagonist" [8]. Again, note that only one of these is strictly positive. We also see a dichotomy in both classification schemes, with scientists leaning towards being either weak and ineffective or having strong, egotistical personalities.

Is there evidence that the prevalence of these negative stereotypes translates into mistrust of real world scientists? An analysis led by George Gerbner of primetime television dramas aired between 1973 and 1983 found that these shows represent the medium through which Americans "encounter science and technology most often" [9]. This study found that, while most individual portrayals of scientists were positive, the percentage of evil scientists was far higher than evil physicians (with 1 in 6 scientists depicted as evil, as opposed to 1 in 20 doctors). Scientists were also more likely to be described as older, odd, unsocial loners, an image summarized by Gerbner as being "somewhat foreboding, touched with a sense of evil, trouble, and peril" [9]. These results agree with more recent data cited by the National Science Board showing that between 2001 and 2016 the percentage of Americans who felt scientists are "apt to be odd and peculiar" rose from 24% to 52% [10] (Fig. 2.1).

Fig. 2.1 Dr. Alexander Thorkel of *Dr. Cyclops* (1940) illustrates the stereotype of the mad scientist (Paramount Pictures, public domain, via Wikimedia Commons)

Reviewing specific examples of fictional scientists demonstrates the staying power of these negative stereotypes (and hence their potential influence on public opinions of both scientists and the science they do). Arguably the most (in)famous fictional scientist is Victor Frankenstein. A victim of his own hubris, Frankenstein's secret experiments not only ultimately lead to his own death, but also the deaths of those he loves. Another helpless scientist in a more recent, yet classic, tale is Mr. Murry, the father in Madeleine L'Engle's 1962 young adult novel *A Wrinkle in Time*. After the government physics experiment in "tessering" (traveling through a fifth dimension) unexpectedly brings him to the alien and hostile world of Camazotz, Murry is held hostage until he is rescued by his children. He calls himself a "wiser and humbler man" and is eager to warn his physicist colleagues that "we know nothing… We're children playing with dynamite. In our mad rush we've plunged into this" [11]. Both a foolish and inhuman scientist, Dr. Felix Hoenikker of Kurt Vonnegut's classic 1963 novel *Cat's Cradle* is emotionally stunted and unable to relate to his beautiful wife. Called odd by all who knew him (including his youngest son), he felt no compunction for his part in creating the first atomic weapons. When asked by the military to create a solution to the logistical problem of carrying out military maneuvers under muddy conditions, he created ice-nine, ignoring the catastrophic implications of a material that instantly solidifies water into ice at room temperature.

The noble scientist is often set in direct opposition with his mad counterpart. For example, director Harry Bromley-Davenport's *Xtro II: The Second Encounter* (1990) features a struggle between mad scientist Dr. Alex Summerfield, director of the Nexus Program, and noble Dr. Ron Shepherd. A previous particle accelerator experiment had created a doorway to a hostile parallel dimension, and as the sole survivor to return from exploring that hell, Shepherd had destroyed the facility to prevent its further use. When only one member of an exploration team sent by the new accelerator returns from the parallel world, Shepherd not only advises against a rescue party, but tries to murder the survivor before she infects the rest of them with the parasitic alien she carries (in homage to the classic film *Alien*). Summerfield, the first to be infected, wants to capture the newborn creature alive, even as it murders most of their ranks. Shepherd banishes his mad colleague to the parallel dimension before he can spread the infection further, thus proving that the mad scientist rarely meets with a good end.

Shepherd's failure to convince his former colleagues of the dangers they face reflects what Margaret A. Weitekamp notes is a frequent impotence in depictions of male scientists. The noble character often "fails to communicate the danger effectively—often because he is so socially removed" or is reduced to a "passive pawn … controlled or co-opted by the military or big business" [12]. Therefore, even when a fictional male scientist tries to act heroically, for example in the role as the lone voice of reason, he is frequently not taken seriously by those around him (often to the detriment of society). While one can argue that this is necessary for the sake of dramatic tension, this emasculation of male scientists propagates a negative stereotype. As Haynes reflects, the bottom line is that "good scientists are clearly in the minority" [7]. The question becomes, is it any better for depictions of women in science (Fig. 2.2)?

2.2   Stereotypes of Female Scientists

Throughout much of the history of Western Culture, science has been viewed as primarily the domain of men, with this stereotype being especially true in the physical sciences. While women scientists are thankfully no longer seen as odd exceptions, the statistics remind us that women in the physical sciences are far from reaching parity with men. For example, in 1975 only 5% of the

Fig. 2.2 The evil Director-General of CERN murders a graduate student in order to hide his responsibility for a zombie outbreak at the facility in *Decay* (2012) (H2ZZ Productions, CC-BY-NC; screen capture by author) (Public domain)

Ph.D.s in physics awarded in America were earned by women; this percentage was 18% in 2016 (down from a high of 23% in 2000 and 2003) [13]. A meta-analysis by three researchers at the University of Melbourne of 36 million authors published in over 6000 STEM (Science, Technology, Engineering, and Mathematics) and medical journals over the past 15 years showed that, at the rate things are improving for women in terms of publications, it will be over 250 years before the percentage of female to male senior authors in physics reaches about 45% [14]. It should also be noted that in 2018 Donna Strickland became only the third woman in history (and the first in 55 years) to win a Nobel Prize in Physics, and in 2016 Fabiola Gianotti became the first woman Director-General of CERN. The reasons are legion for the continued physics gender gap and the lethargy with which it is closing, and reflect biases in both the educational system and employment culture [15]. There are also long-standing issues of sexual harassment and discrimination, which are now increasingly called-out as inappropriate by the larger physics community. A recent high profile example is CERN's condemnation and suspension of particle physicist Alessandro Strumia for making misogynist comments during his presentation at a CERN-sponsored workshop on High Energy Theory and Gender [16]. There is admittedly still a long way to go, but the tide has slowly turned (Fig. 2.3).

Stereotypical images of scientists become ingrained at a young age, as demonstrated in studies of children's drawings. While illustrations of female scientists have become more common in recent decades, children seem to retreat to assuming most scientists are male as they get older [17]. In addition, Jocelyn

Fig. 2.3 The current Director-General of CERN, Fabiola Gianotti, alongside the ATLAS experiment (CERN, CC BY-SA 4.0, via Wikimedia Commons)

Steinke argues that negative stereotypes of scientists (as being socially awkward, unpopular, or unethical, for example) lead girls to be less inclined to consider careers in STEM fields [18]. Representations of female scientists in popular culture therefore have the potential to send powerful messages about the role of women in science and influence the next generation of scientists, for good or evil [19].

In her study of feature films, Eva Flicker found six specific stereotypes of women scientists: the old maid who is married to her work (until she abandons her science and reclaims her femininity through her love for a man); the male woman (a middling, asexual scientist who relies on her assertiveness to survive in an all-male environment); the naïve expert (ethical, good-looking, but ineffective); the evil plotter (an attractive, self-absorbed vixen with questionable morals who wields her sexuality as a weapon); the daughter/assistant (whose character is defined only through her relationship with a male scientist); and the lonely heroine (a strong, competent, ethical scientist and who can simultaneously be feminine but who still requires a male mentor to be successful) [1]. Interestingly Flicker did not encounter female examples of the classic mad scientist working alone in a secret laboratory on dangerous projects. While this may be strictly true for her sample, depictions of fictional female scientists working on particle accelerator related projects are often hardly positive (Fig. 2.4).

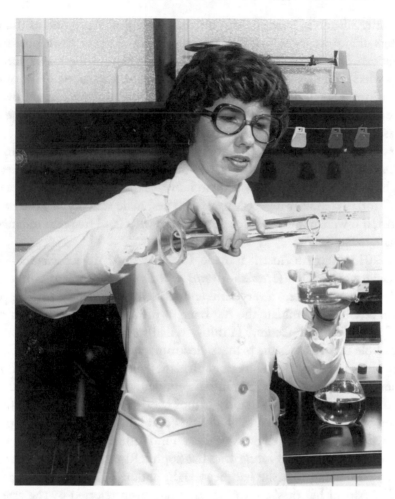

Fig. 2.4 Physical chemist Barbara Askins, the first individual woman to receive the National Inventor of the Year Award, is unfortunately depicted as the stereotype of the "male woman" scientist in this photograph (NASA, Public domain, via Wikimedia Commons)

Take, for instance, director Ron Howard's 2009 film *Angels & Demons*. Here CERN scientists Vittoria Vestra and Father Silvano Bentivoglio have the electricity temporarily cut to their experiment because colleagues are concerned by their attempt to generate and contain a significant amount of anti-matter at the Large Hadron Collider. Vittoria rejects a plea to wait (it is implied until safety issues can be checked), and threatens to call the Director to complain. After her colleague pleads, "Don't blow us all to heaven", an intentional pun on the religion aspects of the film, their experiment is a success [20]. Silvano is visibly uneasy with their result while Vittoria is exuberant. Silvano is immediately murdered in the theft of one of the three canisters of antimatter, which is afterwards used as a terrorist weapon against the Vatican. While she initially demonstrates agency in the above scene (as well as what might be labelled single-mindedness almost to the point of madness), Vittoria is relegated to the weaker stereotype of the old maid over the course of the film, as she slowly becomes romantically involved with the hero, Professor Robert Langdon. She also plays the role of the naïve expert, noting in her admission to Langdon that "the worst thing we thought would happen was that our work would fall into the hands of the energy companies. We thought we could change the world. So naïve" [20]. A case could also be made that Vittoria's character reflects one of Haynes' male stereotypes, the helpless scientist, if it were not for her close association with (and frequent reliance on) the male Langdon.

The 2013 German-Austrian made for television project *Heroes—When Your Country Needs You* (*Helden—Wenn dein Land dich braucht*), directed by Hansjörg Thurn, features a very different depiction of a female particle physicist. An attempt to simulate the Big Bang at the so-called "God Machine" at the "Geneva Research Center" (a thinly veiled stand-in for the LHC) creates a black hole that changes the planet's electromagnetic and gravitational fields. The world is in chaos as satellites and airplanes fall from the sky, the communication grid collapses, and mountains suddenly appear, thrusting upward from the crust. Czech string theorist Luboš Motl sharply criticized the film on his blog for its bad science and "ludicrous opinions about physics" espoused by the film's fictional religious leaders (and other characters) [21]. The hero(ine) of the film, the young scientist Sophie Ritter, has the dubious satisfaction of knowing that she had been right about the dangers she had predicted as part of her thesis, warnings that had been rejected by the scientific establishment. With the aid of her ex-boyfriend and a young male computer hacker, Sophie makes the dangerous trek to the facility and successfully launches the shutdown codes, saving the world. Sophie clearly plays the role of the noble lone voice of reason (a more masculine stereotype), and defies the more stereotypical lonely heroine because she lacks a male mentor to guide her in the film. She also wards off an attack by the villainous accelerator director (who is sucked into his black hole), succeeding where all her male companions have ultimately failed.

Dr. Wen "Wendy" Mei Li, Herman Wouk's fictional female physicist in *A Hole in Texas*, exhibits stereotypically male confidence as she testifies before a Congressional committee about her country's apparently successful Higgs program. Although she strongly ridicules any notion of a "Boson Bomb" and chastises the U.S. for having lost the opportunity to discover the Higgs particle following the cancellation of the SSC, more attention is paid to her previous love affair with fellow particle physicist Guy Carpenter than her current science. She is described by her former paramour as "an able scientist, not a great mind. Competitive as a racehorse, relentless attention to detail, yet not a fussy perfectionist... She's a born project manager, all in all" [22]. He also recalls that she "isn't gorgeous, never was" [23]. These traits paint Wendy as one of Haynes' male women. However, her tastefully feminine fashion sense is lauded by a female reporter for ABC News on a live broadcast, opining "it beats me, Peter, how a scientist, Chinese at that, can have such style sense" [24]. Throughout her testimony Wendy commands respect and wields power, but behind the scenes she is still the sad, unfulfilled woman Guy had exchanged letters with over the years. Wendy's power is further dissipated when Guy realizes that she has not, in fact, discovered the Higgs boson, but rather a classified U.S. satellite project. In the end, she is revealed to be nothing more than the assistant stereotype, as she largely fails as an independent scientist.

Norman P. Johnson's *God's Spark* also features a less than stellar example of a female particle physicist. The fictional anti-science extremist group Billy Smith's Soldiers of God wipes out the greatest minds in American particle physics in a single blow by causing an avalanche during a planned hike at an annual physics conference at the Aspen Center for Physics in Colorado (Fig. 2.5).

The sole survivor, Dr. Brenda Drake, is sent to CERN as the U.S. government representative, where she finds the facility under the control of the Russians, and the newly upgraded LHC under the direction of Dr. Vladimir Chernov, a stereotypical foolish scientist who is the unwitting pawn of his father's political machinations.

Drake is painted as a second-class physicist from the novel's opening chapters. Her talk at the conference is not on her own original research, but rather on Vladimir's new model extending superstring theory. Her controversial talk is met with laughter, boos, and hisses so loud that the moderator has to restore order. She is the only woman on the fateful hike the next day and it is made clear that the best and brightest physicists died. The hike is also the first of many occasions when she is rescued by Dr. Brad Jorsen, an expert on particle collider computer analysis who had dropped off the grid to live in a Colorado cabin. For example, it is only with Brad's astute realization that she is in peril (and directions as to how to outsmart her would-be assassins) that she narrowly escapes an intentional natural gas explosion at her apartment.

It is also important to note that Drake is only sent as the head of the U.S. physics delegation to CERN because, frankly, there were slim pickings left in the U.S. particle physics community after the avalanche. She is not a particu-

Fig. 2.5 The Aspen Center for Physics, the scene of a mass murder of particle physicists in *God's Spark* (Éamonn Ó Muirí, CC BY 2.0, via Wikimedia Commons)

larly original scientist, and at CERN she works with Vladimir to plow through some details in an extension to his model. The author avoids stereotypes of femininity, instead highlighting her tomboy appearance, and Brad even muses how she fits reasonably well in his spare jeans and flannel shirts after the avalanche. Predictably, they become romantically involved, which gives him further reason to continue saving her from Billy Smith's agents. Drake appears to be a nonassertive (emasculated) male woman in Flicker's schema, as well as occupying the secondary assistant role.

A sharp contrast is drawn between Brenda Drake and Dr. Cecilione "Ceci" Galabeiri from the Technical University of Milan, the CERN Director of Computing. As self-assured as Brenda is self-doubting, Ceci wields her exaggerated sexuality as a weapon, making Vladimir's physicist father visibly uncomfortable with her dyed flaming red hair, make-up, and tight fitting, brightly colored dresses that highlight her hourglass figure. But Ceci is perhaps the most astute of all the fictional CERN scientists, and is not only the first to suspect that the Russians are up to no good, but points out to Brenda several important points in secret Russian papers that shed light on the details of the plot. Unfortunately, she fades into the background at the height of the action. Flamboyant, self-assured, and ethical Ceci does not seem to conform to Flicker's schema; in fact, Flicker's categories don't seem to apply very well to a number of the fictional woman discussed so far. This could be due to particle physics being perceived as even more masculine than physics as a whole [25].

While the female scientists mentioned so far are largely ethical (if not sometimes a tad naïve or ineffective), an interesting counterpoint is Dr. Alicia Butterworth, protagonist of Gregory Benford's 1998 novel *Cosm*. At the very least it is fair to say that her ethics are selective. The characterization is doubly unfortunate in that she is also a person of color, an even greater rarity in particle physics fiction (and fact). While over 22,000 white men earned doctorates in physics in the U.S. between 1973 and 2012, only 66 African American women did so [26]. Visiting Brookhaven from the University of California at Irvine, Alicia and her post-doc, Zak Nguyen, receive a stop order on her experiment to collide uranium nuclei with each other at the real-world Relativistic Heavy Ion Collider or RHIC (at the time of the novel's publication still under construction). A judge demands an additional environmental impact report, including specific information about the radioactive output and how it will be shielded (information apparently missing from Alicia's original report). Alicia is peeved by the public's penchant for tying any experiment involving uranium to nuclear weapons, and faced with a delay that will eat up all the time she has been allotted for her experiment, hands in a safety report with faked computer simulations. Although it is not clearly articulated in the novel, there is another potential reason for the public concerns in the novel. Alicia notes that the Brookhaven public relations team likes to describe RHIC experiments as a "mini Bang", referring to the expected eventual creation of a quark-gluon plasma [27]. The bang that occurs in the novel is instead far from mini, as a vacuum pipe is blown out and Alicia's detector is ruined in the creation of a small, heavy, solid sphere. When Brookhaven decides to stop any further uranium experiments until they figure out what has happened (fearing a lawsuit), Alicia and Zak steal the mysterious sphere and bring it back to UC-Irvine to study.

After Brad, Alicia's graduate student, dies in an accident (burned to death by radiation emitted from the sphere), UC-Irvine's Environmental Health and Safety Office wants to isolate her lab while Brookhaven begins legal action to retrieve what they consider to be their stolen property. Brookhaven unwisely repeats her experiment and there is an even larger explosion, causing millions of dollars in damage and creating a 16 meter wide sphere resting in bedrock. While we will revisit the actual nature of Alicia's sphere in a future chapter, it is clear that Benford portrays her as a slightly foolish or helpless scientist, and while not truly "mad", certainly less than completely ethical. Especially disturbing is the ease with which she fakes the revised safety report, given the importance that the real RHIC and LHC safety reports have played in attempts to assuage public fears, a focus of Chapter 5.

Benford's novel does paint an honest picture of the struggles and discrimination faced by a particle physicist who is not only a woman, but a woman of color. For example, the Brookhaven safety officer derides her by suggesting that her proposal for time on RHIC was only granted because of the "minority scientist points that got added to your group's proposal" [28]. Another realistic problem is pressure placed on her by her university to be a minority mentor and take on additional committee work because she is a woman of color (added responsibilities that take away from her precious research time). Finally, Alicia is particularly frustrated that all of the media articles about her discovery make a point of mentioning her race. But once Alicia returns with her sphere to California, it is Benford's depiction of her relationships with men, and the unequal power balance inherent in them, that reduces her agency. For this reason, we will return to Alicia Butterworth in a future chapter. For the moment, we will turn our attention to further examples of fictional particle physicists, and the stereotypes they reflect.

2.3   Further Case Studies

Two significant changes were made in the transition from Robert J. Sawyer's 1999 novel *Flashforward* to the 2009–2010 ABC television series of the same name. First, the time jump of the 137 second long worldwide blackout (during which people see visions of the future) is compressed from 20 years in the novel down to 6 months in the televised version. Second (and most important for this discussion), the focus of the tv series is the response of everyday people and law enforcement officials in Los Angeles to the "flashforward" rather than what series co-creator David S. Goyer terms "elitist scientists" at CERN [29]. The series still includes at least three scientists, the first two being egotistical womanizer and whiz kid/party boy Simon Campos and guilt-ridden widower Lloyd Simcoe. Our introduction to Campos is on a train, where the inebriated physicist is putting the moves on a beautiful woman by bragging that he knows why the flashforward occurred. When she is understandably skeptical, he directs her to Google phone images for "quantum physicist genius". He asks which image popped up first—him winning an award or dressed only in a lab coat, holding goggles in front of his genitals [30]. The second half of the title of the episode—"Scary Monsters and Super Creeps"—accurately describes Simon. In contrast, socially awkward Lloyd Simcoe is attempting to connect with his autistic son after the accidental death of his estranged wife during the global blackout. He, too, claims to know the cause of the event, believing that it was the result of the experiment (perhaps referenced in the first half of the episode's title) that he and Simon were in charge of at the National Linear Accelerator Project (NLAP). Embodying the stereotypes of the helpless scientist turned noble, Simcoe is haunted by the sense of responsibility he feels for the blackout's 20 million deaths and tries to convince his NLAP bosses that they need to come clean with the public.

Simon, on the other hand, embraces his role as the evil alchemist, but although he has questionable ethics, he is not the true mad scientist of the series. He learns that he, too, was the victim of an international conspiracy that had stolen and twisted his brilliant scientific insights and murdered his father. The cabal's plan is to use the information gained in flashforwards for financial gain. Not only does the unnamed conspiracy use Simcoe and Campos' experiment to generate the global blackout, but had slaughtered the victims of a localized blackout experiment in Somalia in 1991 in order to keep their work secret. Dyson Frost, the scientific brains behind the conspiracy and the Somalia experiment, is the true mad scientist of the series, almost cartoonish in his over the top lack of ethics. For example, he not only puts himself through repeated flashforwards, but had been instrumental in the abuse of patients at the Raven River Psychiatric Hospital in Arizona. Here selected patients known to have superior memories were forced to endure blackouts and report in detail what they had seen of the future. While Dyson is eventually killed, and Simcoe and Campos work to minimize the mayhem caused by a second blackout, the short-lived series ends in a cliffhanger that leaves many mysteries unexplained.

Richard Cox's 2005 novel *The God Particle* raises the bar for hyper-cartoonish mad scientists yet another notch. American businessman Steve Keeley survives a three-story fall in Zurich, thanks to the intervention of a shadowy conspiracy orchestrated by terminally ill mad scientist Karsten Allgäuer, a former Nazi researcher. Steve awakens with nanomachines in his brain, resulting in mysterious powers, especially the ability to directly sense all particle fields, including the Higgs, and manipulate them. Meanwhile Mike McNair, the noble but naïve physicist and director of the SSC clone North Texas Superconducting Super Collider (NTSCC), can't understand why he hasn't found the Higgs boson. It turns out that his colleague and college friend, Larry Adams, an NTSCC computer expert and psychologically unstable serial stalker of beautiful women, has been covering up all the Higgs producing events over his psychotic belief that Mike had stolen the affections of one of his college obsessions. Larry can be considered a simple parody of the odd, unsocial stereotype of the scientist, except that his sexual predation (which should never be depicted in a comic manner in the first place) is depicted as completely serious in the novel. Evil plotter Samantha Aizen is brought in by Karsten Allgäuer to undermine Mike, take the credit for the Higgs, and gain Karsten unlimited access to the accelerator in order to fully tap into Steve's powers. Fortunately Samantha choses ethics over loyalty to her employer and warns Mike to work with Steve to stop the mad scientist. Steve reads Karsten's mind and realizes that three other people had previously died from the nanomachine experiment. Understanding that Karsten is trying to somehow use the collider to guarantee his own immortality (it's all a bit nebulous), Steve sacrifices himself in the act of destroying the accelerator ring, leaving the detectors and buildings intact. It is interesting to note that Mike, the noble in intent, but largely impotent in agency, scientist, is not the active hero; rather it is the novel's Frankenstein's monster, the former businessman Steve, who fulfills that role. As for Larry, he disappears at the end, boarding a bus to Hollywood to continue his stalking in person, a rather disturbing loose end.

The film *The Void* illustrates a prime example of the impotent lone voice attempting to warn against a dangerous experiment, although the role is shared by Dr. Eva Soderstrom and, in a flashback, her late father. Eight years before the film's present, mad scientist Dr. Thomas Abernathy, head of the Filadyne Research Center, ignores the senior Soderstrom's concerns that a test at the company's Luxembourg accelerator is putting his colleagues at risk. As the ground shakes and the temperature rises in the accelerator, Soderstrom pleads with Abernathy to shut down the test and evacuate the facility, but it is too late. Containment is lost and the self-destruct mechanism kills the entire Luxembourg staff. In the present, Eva Soderstrom is determined to not let her father's death have been in vain. She uses information surreptitiously obtained from her boyfriend, Filadyne engineer Steven Price, in her independent calculations of Abernathy's latest experiment. Eva is clearly the more intelligent of this romantic pair, something Steven readily admits. Despite using her sexuality to gain the information she wants, Eva is no evil plotter, in Flicker's definition. Instead, her motivations are saving the world first, and revenge second. She shares her calculations with Dr. Jacob Lazarus, until recently a Filadyne physicist, who has suspiciously packed his bags and prepared to leave town. She explains that rather than working on a "clean, boundless source energy", the experiment that claimed her father's life was instead intent on making miniature bombs with the same power as a nuclear weapon [31]. Despite the fact that Lazarus is bound by his nondisclosure agreement with the company, he admits to Eva that the Atomic Energy Commission is funding the experiment, widening the conspiracy. Lazarus warns Abernathy that the calculations are wrong, and is murdered for his trouble, as the now terminally ill Abernathy is determined to let nothing stand in the way of his last chance to succeed. Abernathy finally realizes the error of his ways when the black hole his experiment creates threatens the entire world, and ultimately gives his life to contain it.

Martin Caidin's novel *Star Bright* (1980) illustrates the point that not all mad scientists begin as such, but are instead helpless scientists who make the conscious decision to ignore warning signs. Similarly, not all noble scientists make a conscious decision to sacrifice themselves for the world, but rather find themselves thrust into that position. Retired scientist Professor Owen Kimberly is summoned by the U.S. government to assist when Star Bright, the fusion project he had helped design, spirals out of control and becomes self-sustaining, ultimately forming a black hole. The mad scientists in this novel are twofold, one metaphorical and one literal. Dr. Richard Clayton is the project's young, genius chief scientist. U.S. president Arthur Whiteson notes that Clayton's superior knowledge of physics "made him a supremely dangerous man, dabbling as he did, with supremely powerful forces" [32]. In fact, Whiteson condemns all the scientists of the project for not considering the possible safety risks and consequences of their actions. Owen is the noble, rational hero who surmises that the experiment has somehow generated a microscopic black hole that will destroy the planet if it is not stopped. Dr. Lawrence Pound, the other chief scientist on the project, literally loses his mind over the course of the novel due to the influence of the black hole. Clayton and Pound also fulfill the tropes of the foolish scientist and helpless scientist of Haynes' analysis, with Russian scientist Vasily Tretyakov accosting the Americans for their apparent lack of safety concerns. He calls them "Children playing with dynamite" and asks if they are "so hungry for energy, so greedy for fame and your dollars, that you would not observe even the most basic rules of control? Of safety?" [33]. Kimberly and Tretyakov devise a plan to fire 45 antitank weapons at once from all directions, hoping to disrupt the black hole's apparent connection with another universe. When the black hole grows faster than anticipated, the scientists' automatic firing mechanism has to be set to manual, and the noble Owen Kimberly dies (something he had tried to avoid) heroically saving the planet.

A sharp contrast can be drawn between Kimberly and of all the scientists in Garfield Reeves-Steven's science fiction-detective novel *Dark Matter* (1990). Brilliant Nobel Prize winning physicist and psychopath Dr. Anthony Cross tortures women by dissecting their living brains, not, as in the case of Hannibal

Lecter, to consume them, but instead slithering his fingers inside their opened skulls while they are still conscious in order to experience the moment of their deaths and through that death come a bit closer to understanding the moment of creation. Cross is protected by his lover, Dr. Charis Neale, a fellow particle physicist and mathematician, and intellectually inferior collaborators (and stereotypically helpless scientists) Dr. Adam Weinstein and Dr. Lee Kwong, who nevertheless share Cross' Nobel Prize. Charis is left out of the prize because only three individuals can share the award and she is considered by many to merely be Cross' assistant (presumably due to her gender). This diminishing of Charis' part in the project brings to mind similar controversies with Antony Hewish's share in the 1974 Nobel Prize in physics for work that was largely done by his then graduate student, Jocelyn Bell, and the role of Rosalind Franklin in Francis Crick and James Watson's work in discovering the structure of DNA. Charis' most important roles are ultimately to keep Anthony from becoming completely unraveled and cleaning up his bloody messes (a rather gruesome extension of Flicker's assistant stereotype).

A private company provides the quartet with significant resources in order to continue their research, especially as it relates to potential military applications. Los Angeles Detective Kate Duvall deduces that Cross is the serial murderer, but when she confronts Neale the physicist offers excuses for his behavior. According to Neale, Cross' victims constitute "acceptable losses" because Cross is a genius in both science and "in manipulating people.... But he's probably the greatest mind we've ever known, so isn't that worth the greatest sacrifice?" [34]. Cross's ultimate scientific triumph is a glove that allows his hand to behave like an uncertain quantum particle/wave so that he can reach into people's bodies and kill them, including a less messy "upgrade" of his crude brain surgery.

While Charis Neale protects her lover and rationalizes his cruelty until the end, Dr. Julia Wilson's loyalties shift over the three seasons of *The Boat*. A particle accelerator accident in Geneva creates a black hole that absorbs about 500 km^3 of central Europe as it falls toward the center of the Earth. The ongoing conspiracy behind the accelerator and the ship's fate is slowly revealed over the course of the series, reaching back into the past to the early development of accelerators. As described in Chapter 1, at the heart of the conspiracy is the Alexandria Project, modeled on a project begun during World War II by an unnamed scientist to prevent the extinction of the human species in the event of a nuclear holocaust. Seven well-equipped ships with hand selected crews of breeding populations are launched to ride out a potential catastrophe as the CERN-inspired ECND (European Commission for Nuclear

Development) prepares to fire up a new experiment at its unnamed supercollider in Geneva (an obvious nod to LHC).

This super-secret conspiracy is initially supported by over a dozen governments, but when the morally corrupt financier Alexander Montes takes on a key role in the project, its humanitarian mission is corrupted to his personal benefit, guaranteeing that he will also survive aboard a secret submarine shadowing the seven ships. The central scientists in the series are Dr. Julia Wilson, Dr. Roberto Schneider (the assumed name of Dr. Roberto Cardeñosa), and several particle physicists at the ill-fated ECND accelerator. Julia is not only the *The Polar Star*'s brilliant medical doctor, but had worked on the accelerator's B Protocol for assuring the survival of the human species in the case of a disaster, drawing upon her expertise in physics and biochemistry. The conspiracy deepens under Montes' influence, and Julia regrets becoming involved; she is afterwards threatened into continuing her part in the project. After the planet-changing cataclysm, she slowly divulges information to Captain Ricardo Montero and eventually works with him against her former employers, knowing that she is putting them all in jeopardy due to the presence of a murderous conspirator on the ship. Julia is painted as an ethical scientist who gets in over her head as part of a project that she initially believes has an honorable intention. She also explains to Ricardo that the odds of the accident occurring were only one in a million (or one in a billion—the series isn't consistent on this point) and that the secrecy is simply necessary to avoid causing a panic. In terms of Flicker's classification of female scientists, while Julia is certainly attractive (and her love affairs, first with the much younger Ulises Garmendia and ultimately with the Captain, play central roles in the series), she is neither the evil plotter nor ineffective naïve expert. The trope of lonely heroine is also not completely relevant, as she is the de facto chief scientist of the *The Polar Star* and has no male mentor. Julia Wilson is therefore a rather complex character whose ethics evolve over the course of the series (Fig. 2.6).

The primary male scientist of the series is Roberto Cardeñosa, only truly seen in flashback. Initially an egotistical but brilliant particle physicist, Harvard professor, and author of the book *Particle Acceleration and Black Holes*, he is hired by ECND to work on the Geneva supercollider project. He plays a central role in creating the B Protocol of the Alexandria Project after the ECND accelerator discovers how to create matter. Weighing the potential benefits of practical applications against the inherent dangers of this discovery, as well as the need for utter secrecy, Roberto cuts ties with both his beloved older sister Salomé and his fiancée, changing his last name to Schneider. As Montes' influence turns the project more and more sinister, Roberto admits in a cellphone confession that the work he had "done to make a better world

Fig. 2.6 Actress Irene Montalà (Dr. Julia Wilson) from *El Barco* (*The Boat*) (Estrella Damm, CC BY 2.0, via Wikimedia Commons)

now only serves to destroy it" [35]. He attempts to warn his superiors that the experiment is too dangerous to proceed, but when they push on anyway, he engineers a mysterious debilitating chronic respiratory disease with which he infects many of the Alexandria Project personnel. After hiding information on the cure in a red folder on *The Polar Star*, he blackmails his superiors with it in a desperate attempt to stop the collider experiment. In return, they attempt to murder him through drowning, leaving him brain damaged. Afterwards he takes on the name Bubble and lives on the *The Polar Star*, taken care of by his sister Salomé, the ship's cook, who honors his pre-accident request to never reveal that they are related (for her safety). While he does not recall any of his work for the ECND, he knows that something called a particle accelerator was involved and, more importantly, he remembers hiding the folder.

Roberto is undoubtedly the most complex of all the characters in this brilliant ensemble cast, and arguably the most sympathetic, given his mental and emotional challenges. In his role as part of the ECND collider experiment he would be termed equal parts evil alchemist and helpless scientist. He cannot be called a mad scientist because his intentions were not only noble, but he tried to stop the experiment when he realized it was dangerous. While creating a debilitating (but not deadly) disease and withholding the potential cure can hardly be termed ethical on face value, given that his intention was to prevent the destruction of the planet perhaps the ends justify the means. At the end of the series he produces the cure for the surviving Alexandria Project personnel in order to secure the safety of *The Polar Star*, and despite his continued mental challenges is reunited with his fiancée, finding greater happiness in his simple life as Bubble than he ever found as the brilliant particle physicist Roberto.

The lesson of Bubble's transformation is juxtaposed against the fate of the majority of the ECND physicists, especially Philippe. Involved in a forbidden, clandestine romantic relationship with Julia before the cataclysm, Philippe is seen as the narrator of the project's "in the case of a disaster" instructional videos and in flashback, setting up his role as an important player in the project. In the first episode Philippe and an unnamed colleague bring Julia to her post on *The Polar Star* in an ominous black van, as befits such a serious conspiracy. While Philippe tries to exude confidence as to the success of the scheduled accelerator experiment, Julia senses his obvious anxieties, and he is forced to lie about the experiment's status. Most importantly, he does not disclose a serious incident with the magnets (which presumably later causes the cataclysm). The degree to which the ECND scientists are mad scientists rather than evil alchemists or helpless scientists is open to debate, and is based on their knowledge of the potential for disaster. Regardless of where the viewer stands on their classification, the overall message of the series is clear—particle physics is dangerous, safety protocols are unreliable, and ethical scientists are the clear underdogs against nefarious business and financial interests.

Safety issues and the intentions of particle accelerator physicists are also important in the ongoing American series *The Flash*. As a superhero drama, there is a clear need for heroes and villains, but in *The Flash* these roles are unusually complicated, as a number of characters are definitely not who they initially seem to be. There are convoluted plotlines based on time travel, with multiple time lines that shift as characters change the past, as well as motion between parallel universes. Not only can you meet a version of yourself from another time or universe, but you can actually take the place of your other "you". An important example is Harry Wells, a scientist from the parallel world dubbed Earth 2, who temporarily takes the place of the "Harrison Wells" of our Earth (Earth 1). This latter character is, in turn, actually the twenty-second century mad scientist Eobard Thawne masquerading as the original Earth 1 scientist he had murdered years earlier. Are you confused yet?

The basic plotline of the series involves the S.T.A.R. (Scientific and Technological Advanced Research) Laboratories synchrotron whose Earth 1 version becomes nicknamed the Pipeline. In the original timeline, S.T.A.R. labs is the dream of the noble and humble scientist Harrison Wells (the murdered original) and his scientist wife Tess Morgan, envisioned as an independent lab dedicated to progress in science through service to humanity without connections to governments or industries. They realize their dream by 2020, but an accident with dark matter creates the "speedster" called the Flash (so-named because he can run at incredible speeds) and other mutated meta-humans. In the future Eobard Thawne becomes obsessed with the Flash, and reverse engineers the accelerator accident in order to become a speedster himself. Through further experiments with the so-called Speed Force he discovers time travel and travels back to the year 2000 in order to assume Wells' identity and hasten the development of the S.T.A.R. accelerator. In this new timeline, the synchrotron is intentionally designed to "fail" as a particle physics experiment (or as a source of clean energy, its advertised purpose), and when it is turned on for the first time on December 11, 2013, a carefully planned explosion kills 17 and infects thousands of people with dark matter, most notably mild mannered police forensic scientist Barry Allen. After the experiment the mad scientist Wells/Thawne feigns being confined to a wheelchair (to disguise his true power and gain the sympathy of others) and helps Barry hone his new superpowers. With the aid of his unsuspecting assistants, bioengineer Dr. Caitlin Snow and mechanical engineer Cisco Ramon, Wells/Thawne pretends to work for the common good by turning the RF cavities of the accelerator into prison cells for the ill-intentioned meta-humans. Unbeknownst to his colleagues he is simultaneously repairing the accelerator in order to return home to his time with newly enhanced speedster powers.

In contrast, the Earth 2 version of S.T.A.R. Labs is far more technologically advanced than our Earth's version, founded in 1991 by that world's Harrison "Harry" Wells, a cynical and sarcastic but not "mad" scientist who had built his accelerator for the benefit of his world. When his accelerator unintentionally explodes, he directs the explosion underground, trying to limit human exposure to the dark matter. Instead, the explosion infects an asylum for the criminally insane, including Hunter Zolomon, a convicted serial killer. Zolomon becomes a speedster known as Zoom, using his powers to terrorize both his world and ours. Harry Wells eventually accepts his role as the helpless scientist, taking responsibility for the creation of Zoom and other immoral meta-humans whom Zoom turns into his army of evil. In response, Harry evolves into a noble scientist, working alongside Barry, Caitlin, and Cisco to use his scientific creativity and technological prowess in the fight against Zoom. As for

Zoom, he fully embraces the role of mad scientist, experimenting with various drugs in order to increase his speed, leading to physical harm and the need to seek out other speedsters in order to sustain himself. He kidnaps a speedster from yet another Earth, Jay Garrick, and takes on his identity on Earth 2, playing a dual Jekyll and Hyde role as the heroic Jay/Crimson Comet Flash and the nefarious Zoom. Among those who fall for this ruse are Barry's team, most especially Caitlin Snow, who falls in love with the Jay persona.

Recent events had rendered the compassionate and intelligent Caitlin especially susceptible to Jay's plot to manipulate her. She had been devastated by the apparent death of her beloved fiancé Ronnie Raymond, a S.T.A.R. labs engineer, in Wells/Thawne's intentional explosion, only to be later further traumatized when Ronnie miraculously reappears but sacrifices himself to save the world not long after their wedding. She blames herself her Ronnie's death, and leaves S.T.A.R. Labs for a time to work for Mercury Labs, where she is mentored by female scientist Dr. Christine "Tina" McGee (a valuable experience after her tenure at the all-male S.T.A.R. facility). Caitlin's locus in the Flicker classification system is unclear. She is hardly ineffective, although she is certainly naïve and easily manipulated by the Jay side of Hunter; while she is romantically involved with several characters, her sexuality is only weaponized when she uses Jay's infatuation with her to lure him into Barry's trap. However, it is important to understand that this was only after Caitlin had been kidnapped and imprisoned by Jay/Zoom, demonstrating her inherent inner strength.[1] She clearly craves mentorship, and finds it not only in Tina McGee but Earth 2's Harry Wells, who compliments her before returning to his Earth with the parting phrase "You're a tremendous scientist, but—you're an even better person" [36]. The characterization of Caitlin Snow as a female scientist is therefore complicated, and although it includes positive aspects, taken as a whole feeds negative stereotypes of women, especially that of a lonely woman as easily manipulated by an attractive man.[2]

Other stereotypical scientists appear in the first two seasons of the series, most especially the titular character, the "Flash" Barry Allen. An intelligent consummate nerd, he is depicted in the early episodes of the series as socially awkward, especially in his inability to articulate his feelings for his best friend, Iris West. In the inaugural episode of the series he is teased by Iris for wanting

---

[1] Caitlin's treatment by Jay/Zoom/Hunter is certainly problematic from a feminist point of view. In later seasons she appears to suffer from PTSD, but this is beyond the scope of this analysis.

[2] A brief mention should also be made of Harry Wells' daughter Jesse, dubbed Jesse Quick by her father due to her innate intelligence and quick thinking. She proves to be a valuable member of the scientific team during Season 2, despite the youthful recklessness and rebellion against her father she sometimes exhibits.

to attend the ceremonial startup of the S.T.A.R. accelerator, jokingly calling it Barry's "sad little nerdy dream". When Barry gushes about the scientific importance of the accelerator, she offers "You gotta get yourself a girlfriend", an ironic statement given Barry's feelings for her [37]. It should be noted that Barry and Iris only become romantically involved once he becomes a superhero and thus gains significant physical prowess, hearkening back to Weitekamp's argument concerning the common trope of impotence (a lack of manliness) in fictional scientists. His blind obsession with having witnessed the death of his mother at the hand of Thawne when he was a child causes him to act impulsively and exhibit extreme hubris, going so far as to intentionally change the timeline to save his mother despite understanding that it will alter the lives of everyone else on the planet. It can be argued that Barry's attachment to his mother is depicted as unhealthy and adds to the emotional impotence of the character, belying the strength and masculinity emphasized by a super-hero persona.

Finally, two young male assistants to Wells/Thawne, mechanical engineer Cisco Ramon and Hartley Rathaway, should be mentioned. The egotistical and brilliant whiz-kid Rathaway, who openly refers to himself as the "Chosen One", is the protégée of Wells/Thawne at S.T.A.R. Labs before the accelerator accident. Described as having a "challenging personality" by the more politically correct Caitlin, Cisco, whom Rathaway treats with open derision when the young engineer joins the lab, dubs his former tormentor "mostly a jerk. But, every once in a while, he could be a dick" [38]. Despite his grating personality (which is unchanged after he develops meta-human powers) Hartley correctly predicts that the accelerator could explode, and therefore, perhaps unwittingly, plays the role of the noble scientist before he is fired by Wells/Thawne (who, remember, fully intended for the accelerator to explode). In contrast is the equally intelligent, well-liked, and affable Cisco Ramon. Armed with a limitless supply of geeky popular culture references (ranging from *Star Trek* and *Star Wars* to *Jaws, Young Frankenstein*, and *The Walking Dead*), Cisco's clever quips include his talent for coming up with catchy nicknames for various meta-humans. Cisco's approachability and lovable nerdiness allow him to occupy the role of avatar for the audience within the series, the reluctant meta-human who is fallible, vulnerable, and, well, human, a welcome change from the more godlike Wells (in all his versions).

Taken as a whole, what do these examples teach us? Evil, foolish, helpless, and noble scientists certainly abound in particle physics media. Interestingly it can be argued that many of the female scientists are more closely aligned with male stereotypes than typically female ones, as previously noted perhaps a reflection of particle physics' hyper-masculine reputation. This could also be interpreted more positively, in portraying female particle physicists as more independent and self-sufficient than women in other fields of science. Whatever interpretation we take, we should remember that depictions of particle physicists behaving badly in popular culture have the potential to affect the public's perception of real world scientists. But as we explore in the next chapter, the public's preexisting misconceptions and inherent uneasiness at several aspects of accelerator experiments (especially concepts of high energy, modelling the Big Bang, and producing radiation), as well as common anxieties over natural disasters and a general fear of astrophysical events (cosmophobia), also have an important part to play in generating mistrust of particle physics.

References

1. E. Flicker, Between brains and breasts – women scientists in fiction film. Public Underst. Sci. **12**, 308–310 (2003)
2. C. Sagan, *The Demon-Haunted World* (Ballantine Books, New York, 1997), p. 11
3. J.R. Wheeler (dir.), The Krone Experiment (2005)
4. J. Cramer, *Einstein's Bridge* (Avon Books, New York, 1997), p. 19
5. J. Cramer, *Einstein's Bridge* (Avon Books, New York, 1997), p. 85
6. J. Cramer, *Einstein's Bridge* (Avon Books, New York, 1997), p. 127
7. R. Haynes, From alchemy to artificial intelligence: stereotypes of the scientist in western literature. Public Underst. Sci. **12**, 244 (2003)
8. K. Grazier, S. Cass, *Hollyweird Science* (Springer, Cham, 2015), p. 79
9. G. Gerbner, Science on television: how it affects public conceptions. Issues Sci. Technol. **3**(3), 110–112 (1987)
10. National Science Board, *Science & Engineering Indicators 2018* (National Science Foundation, Alexandria, 2018), pp. 7–66
11. M. L'Engle, *A Wrinkle in Time* (Square Fish, New York, 2007), p. 185
12. M.A. Weitekamp, The image of scientists in *The Big Bang Theory*. Phys. Today **70**(1), 43 (2017)
13. Percent of Physics Bachelors and Ph.D.s Earned by Women, Classes of 1975 Through 2016, American Institute of Physics, https://www.aip.org/statistics/data-graphics/percent-physics-bachelors-and-phds-earned-women-classes-1975-through-2016
14. L. Holman, D. Stuart-Fox, C.E. Hauser, The gender gap in science: how long until women are equally represented? PLoS Biol. **16**(4), 6 (2018)
15. Why so Few? Women in Science, Technology, and Mathematics, American Association of University Women, https://www.aauw.org/resource/why-so-few-women-in-science-technology-engineering-mathematics/

16. CERN Scientist Alessandro Strumia Suspended after Comments, BBC News, https://www.bbc.com/news/world-europe-45709205
17. D.I. Miller, K.M. Nolla, A.H. Eagly, D.H. Uttal, The development of children's gender-science stereotypes: a meta-analysis of 5 decades of U.S. draw-A-scientist studies. Child Dev., 7–9 (2018). https://doi.org/10.1111/cdev.13039
18. J. Steinke, Cultural representations of gender and science: portrayals of female scientists and engineers in popular films. Sci. Commun. **27**, 36 (2005)
19. A.L. Carlson (ed.), *Women in STEM on Television: Critical Essays* (McFarland & Company, Jefferson, NC, 2018), pp. 4–5
20. R. Howard (dir.), Angels & Demons, Sony Pictures (2009)
21. L. Motl, LHC Black Hole: A Catastrophic Movie, The Reference Frame, https://motls.blogspot.com/2014/04/lhc-black-hole-catastrophic-movie.html
22. H. Wouk, *A Hole in Texas* (Little, Brown and Co, New York, 2004), p. 96
23. H. Wouk, *A Hole in Texas* (Little, Brown and Co, New York, 2004), p. 107
24. H. Wouk, *A Hole in Texas* (Little, Brown and Co, New York, 2004), p. 216
25. S. Traweek, *Beamtimes and Lifetimes* (Harvard University Press, Cambridge, 1988), p. 16, 79
26. J. Levine, These Shocking Charts Show How Hard It Is for Black Women in Science, Mic, https://mic.com/articles/122185/stem-black-women
27. G. Benford, *Cosm* (Orbit, London, 1998), p. 17
28. G. Benford, *Cosm* (Orbit, London, 1998), p. 5
29. "Architects of Destiny", Featurette, FlashForward: The Complete Series, DVD, Buena Vista Home Entertainment (2010)
30. S. Hoffman, Q. Peeples (script), Scary Monsters and Super Creeps, FlashForward, season 1 (2009)
31. G. Shilton (dir.), The Void, Lions Gate Entertainment (2001)
32. M. Caidin, *Star Bright* (Bantam Books, New York, 1980), p. 15
33. M. Caidin, *Star Bright* (Bantam Books, New York, 1980), p. 124
34. G. Reeves-Stevens, *Dark Matter* (Bantam Books, New York, 1990), p. 340
35. E.M. Lobato, G. Navas (script), Fukushima Blues, The Boat, season 3 (2012)
36. A. Helbing, T. Helbing (script), The Race of His Life, The Flash, season 2 (2016)
37. A. Kreisberg, G. Johns (script), Pilot, The Flash, season 1 (2014)
38. A. Schapker, B. Eikmeier (script), The Sound and the Fury, The Flash, season 1 (2015)

3

The End Is Nigh: Fundamental Particle Fears

3.1 Things that Go Boom

John Ringo's 2005 novel *Into the Looking Glass* begins with an explosion equivalent to 60 kilotons of TNT, marking the accidental destruction of a university particle accelerator in Florida. The preface to Dan Brown's novel *Angels & Demons* makes a point of describing for the reader just how powerful antimatter is, with the reaction of one gram of antimatter and the same amount of matter releasing energy on par with a small atomic weapon. In Martin Caiden's novel *Star Bright* a fusion experiment creates a mini star that evolves into a miniature black hole. Russian scientist Vasily Tretyakov admits that the Russians had previously developed a superbomb using similar technology to Star Bright that had resulted in a flood of unknown radiation that created "a tornado of naked electrical force" [1]. The Russians aid the American scientists in trying to contain Star Bright, but earthquakes intensify and the strange electromagnetic effects caused by the unknown radiation lead to widespread irrational behavior.

Speaking of irrational, in his ultimate evil plot to destroy all universes except our own (which he intends to rule), *The Flash's* evil speedster and mad scientist Zoom steals and coopts Dr. Tina McGee's Magnetar, an artificial pulsar described as a "power amplifier with a highly magnetized, dense rotating core that can be easily weaponized" [2]. A real pulsar, is, indeed, highly dense (having twice the mass of the sun but compressed into a ball about 15 miles wide, or about a hundred trillion times denser than lead), rapidly rotating (up to hundreds of rotations per second), and does possess an intense magnetic field a trillion times stronger than that of our planet. A pulsar forms

© Springer Nature Switzerland AG 2019
K. Larsen, *Particle Panic!*, Science and Fiction,
https://doi.org/10.1007/978-3-030-12206-5_3

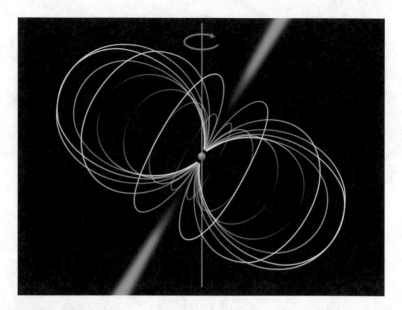

Fig. 3.1 Schematic diagram of a pulsar, highlighting its magnetic field lines and the beams of synchrotron radiation emanating from the magnetic poles (Users Mysid and Jm smits, CC BY-SA, via Wikimedia Commons)

when a star larger than our sun explodes as a supernova, with the core (now converted into a dense ball of neutrons) collapsing to form a corpse called a neutron star. This gargantuan explosion (liberating about as much energy in its death as our sun will over its entire lifetime) is heralded by a shower of neutrinos, screaming its warning call throughout the cosmos at nearly the speed of light. But the property that turns a neutron star into a pulsar is the release of synchrotron radiation. The most powerfully magnetic pulsars (with fields 1000 times greater than normal) are called magnetars (Fig. 3.1).

Described in these ways, accelerators, antimatter, fusion, and pulsars do, indeed, sound scary, with their high energy, fierce magnetic fields, and radiation. Importantly, a pulsar also sounds suspiciously like a particle accelerator, which in some ways it is. Popular media can only capitalize on the public's fears concerning particle accelerators if the public does, indeed, harbors such concerns. We will now put the nature and source of some of these concerns under the microscope, and begin to consider how we can combat them, especially in the age of the Internet, with its instant access to both factual and fictional information. It is also important to recognize an important facet of the current American mindset, namely its morbid fascination with the end of the world. The current popularity of zombies and dystopias (as demonstrated by the high ratings of *The Walking Dead*, *The Handmaid's Tale*, and other media) did not develop in a vacuum. Rather, as Daniel Wojcik explains, apocalyptic thinking is widespread among American evangelical Christians [3]. For example, a 2010 Pew Research Center poll found that 41% of Americans believe that the Second Coming of Christ will occur by the year 2050 [4]. Such ideas clearly feed into the American doomsday prepper movement in which individuals are building and stocking underground bunkers with the intent to ride out the coming apocalypse. The number of American preppers is not precisely known, but a 2012 poll suggests that over a quarter of Americans personally know a prepper [5]. These apocalyptic preoccupations also help explain the widespread interest, especially on social media, in end of the world predictions, including the infamous Mayan calendar December 21, 2012 apocalypse frenzy. While it was a worldwide phenomenon, it found its greatest audience in the already apocalypse-attentive America, as reflected in director Roland Emmerich's 2009 American blockbuster film *2012*.

A far lower budget fictionalization summarizing many of the most popular end of the world scenarios is *End:Day*, a 2005 BBC docudrama directed by Gareth Edwards. The program follows a series of parallel scenarios affecting one day in the life of Dr. Ron Howell (blending original drama footage with interviews of scientists from previous BBC science programs). Howell, the deputy director of the world's largest particle accelerator at the TBM facility near New York (probably based on the RHIC collider at Brookhaven National Laboratory on Long Island), attempts to fly to New York from England for the opening of the facility. He finds his travel plans thwarted by a variety of end-of-the world scenarios: a mega tsunami devastating New York City, a killer asteroid hitting Germany, the Yellowstone supervolcano erupting, and a highly contagious virus. Howell makes it to New York in the final segment, because the doomsday scenario is, in fact, his accelerator creating a black hole (called "one of the most destructive forces in the universe"), a strangelet (having "the power to devour the earth"), or both (as both scenarios are mentioned in the fake newscasts) [6]. As Howell turns the key to start up the experiment, the accelerator explodes, incinerating not only the scientists, but also a large group of protestors outside the facility. Freak vortex-like storms disrupt Paris, shredding airplanes as the International Space Station helplessly watches. The program ends with an explanation by particle physicist Frank Close that because the Big Bang did not destroy the universe, neither will experiments attempting to recreate some of its properties. In Close's words, "it makes great science fiction, great entertainment, great television, but on this, we can sleep easy" [6]. Given the powerful images the viewer has just seen, it is quite possible that Close's assurances will either be ignored or viewed with suspicion. While words have the power to comfort, they also have tremendous potential to confuse and create fear. This chapter focuses on specific examples of physics related terms and concepts that accelerate public fears due to their misuse and misunderstanding.

3.2   High-Energy Collisions and Mini Big Bangs

In Chapter 1 units of energy were casually bandied about with little explanation, awaiting the arrival of this chapter. There are three important points to understand concerning particle physicists and energy. The first is that the common unit used, the electron volt or eV, is so incredibly tiny that in order to describe any meaningful reactions we need to speak of millions of eV or MeV, billions of eV or GeV, and trillions of eV or TeV. Second, Einstein taught us that mass can be converted into energy and vice versa through the most famous equation in all of physics, $E = mc^2$. The "c" in the formula refers to the speed of light in a vacuum, 300,000 kilometers per second or 186,000 miles per second, which is the ultimate speed limit of the universe. Particles in accelerators travel at more than 99% c but can never reach 100%. Since the speed of light is a large number, and a large number squared is an even bigger number, the rate of exchange between mass and energy is quite high. In other words, a small amount of mass can be converted into a tremendous amount of energy. This is why particle physicists often refer to the mass of a particle like the Higgs boson in terms of its energy equivalency. Third, mass/energy can neither be created nor destroyed, only transferred into different forms of each other. Putting this together, we get to the heart of the matter. We need to accelerate particles to more than 99% the speed of light so that when we smack them together their energy of motion (kinetic energy) plus their mass/energy equivalents equal enough energy to make even heavier particles that we want to study. The total amount of mass/energy before the collision equals the total amount of mass/energy after the collision. In the LHC the collisions have total energies of up to 13 TeV, which allows particle physicists to generate a number of heavy (and very interesting) particles. A flying mosquito has a kinetic energy of about 1 TeV, so stated in this way 13 TeV might not seem like a lot of energy. The key is that the LHC is packing this energy into two microscopic protons, not a very much larger insect [7].

Despite these facts, CERN is used to members of the press and public asking if the record breaking collisions of the LHC are dangerous, and include an answer to the question "Are the LHC collisions dangerous?" in their FAQ [8]. The crux of their fear-dissuading argument is that the energies produced in the LHC are dwarfed by energies produced in nature in the form of the energies of so-called cosmic rays. They argue that since these "rays" do not harm the Earth or any other astronomical body, there is no reason to fear human-caused experiments that merely mimic (at lower energy) this natural phenomenon. The admittedly mysterious origin of cosmic rays (the term "ray" itself invoking visions of harmful ray guns found across science fiction media) itself detracts from the comfort of the explanation. Discovered in 1912, these fast moving particles (mainly protons and various atomic nuclei) collide with our atmosphere with energies up to a billion times greater than the collisions achieved in the LHC. Identifying the ultimate cosmic source of these tiny projectiles (besides the sun, supernovae, and most recently the overly-energetic cores of some galaxies [9]) has proven difficult for astronomers, adding to their mystique. In addition, cosmic rays are the primary suspect in astronauts' seeing bright flashes of light, even when their eyes have been closed [10]. Therefore, while comparisons between particle accelerator experiments and the most energetic natural collisions are meant to calm fears, they may actually have the opposite effect (Fig. 3.2).

Likewise, comparisons between particle collisions at the LHC and events that naturally occurred in the early history of our universe can also fall flat. Recall from the last chapter that in Gregory Benford's *Cosm*, particle physicist Alicia Butterworth muses about how facility publicists refer to collisions at

Fig. 3.2 Launch of the CREAM III (Cosmic Ray Energetics and Mass) balloon flight. Instruments measured cosmic ray interactions with our atmosphere from an altitude of about 25 miles (Eun-Suk Seo, CC0, via Wikimedia Commons)

RHIC as a "mini Bang" (in contrast to the Big Bang) [11]. Indeed, this language is widespread in popularizations, and can lead to concerns that physicists are playing God. Many of these "mini Bangs" focus on recreating the quark-gluon plasma of the early universe, as described in Chapter 1, and it must be emphasized that colliding a controlled number of particles together at the same energies that once filled the universe as a whole is not recreating the Big Bang, but merely modelling, on the smallest scale, collisions between these individual particles similar to what they might have experienced way back in the early universe. It is precisely this type of experiment that initially appears to be to blame for the mayhem and destruction caused by the global blackout in the novel *Flashforward*. Therefore, while the potential for high-energy collisions causes a particle physicist's heart to race with anticipation, it can also generate heart pounding fear in members of the public.

3.3   Fears of Radiation

The mysterious sphere created in Dr. Alicia Butterworth's uranium collision experiment turns deadly when a graduate student receives a lethal dose of ultraviolet radiation from it. Ironically, this death allows theoretical physicist Dr. Max Jalon to determine that what appears as a now bowling ball sized object in our universe is actually the birth of a separate child universe (as explained in Chapter 7). The radiation is nothing more nefarious than what happened in the evolution of our own universe when it was a few 100,000 years old (the important difference being that there were no fragile human observers present when our own universe was this inhospitable). This episode from Gregory Benford's novel *Cosm* illustrates the potential harm that can be caused when human tissue interacts with high-energy electromagnetic radiation. Note that I could have easily exchanged the word "light" for "electromagnetic radiation" in the last sentence, because the two terms refer to the same phenomenon. However, to a nonscientist, "light" refers to what we can see with our eyes (what physicists term "visible light"), while "radiation" is thought by the public to automatically refer to something harmful. While ultraviolet wavelengths, no matter what you call them, do pose a danger to human tissue (which is why sunscreen exists), benign violet wavelengths can also be termed "radiation" (as can yellow, red, infrared, radio, and many other wavelengths of light). To the average person on the street, radiation is a code name for danger, so all a novelist or screenwriter has to do is call something a kind of radiation and the audience automatically understands that it is meant to be harmful, even if they don't understand why or how.

J. Bryan Lowder reflects that during his stint as a science writing intern "radiation" was a term that his colleagues were "cautioned never to use in our public articles" specifically because of this fear response on the part of general readers [12]. The root cause of radiation-phobia is pretty clear. As Susanne Neumann explains, most appearances of the word in media and common conversation relate specifically to nuclear radiation, especially when discussing nuclear weapons or nuclear waste (both of which even nonscientists realize are rather unhealthy) [13]. From haunting images of Hiroshima survivors to reports of increased cancers rates following the Chernobyl nuclear power plant accident, our experiences with "radiation" are rather negative. This colloquial use of the term leads to misconceptions that all radiation is not only harmful but also human-made. From *Godzilla* and *The Incredible Hulk* to *The Attack of the 50 Foot Woman* and *The Night of the Living Dead*, the consumer of popular culture is bombarded with countless representations of the simple equality radiation = bad. It would be scientifically correct to distinguish between harmful *ionizing radiation* (such as X-rays, gamma rays and high-energy particles) that carries sufficient energy to rip electrons off atoms and *non-ionizing radiation* such as visible light, but the additional use of buzz words might only further confuse the public (Fig. 3.3).

It is an unavoidable fact that particle accelerators produce ionizing radiation; for example, recall the synchrotron radiation resulting from the acceleration of charged particles around a circular track, and that the betatron was used to produce X-rays. The public relations office of CERN takes special care to address the safety of both their employees and the general public in regards to radiation, noting that having the accelerator tunnel 100 meters underground provides a natural shield. The result of all the CERN safety measures is that the average citizen of the region receives 240 times more ionizing radiation from the natural environment (for example, from cosmic rays and radioactive minerals) than from CERN [8].

Science fiction writers have both greatly exaggerated the dangers from particle accelerator radiation (as in the meta-mutants of *The Flash*) as well as conveniently ignored its presence (as the plot dictates). For example, in the German-Austrian made-for TV movie *Heroes* (*Helden*), kindergarten children are given a tour of a facility based on CERN and watch the experiment from a huge glass window overlooking the colliding beams. Even more unrealistic are the lab coat wearing scientists seen standing directly underneath the operating beam. Likewise, in 2009s *GI Joe: The Rise of Cobra*, directed by Stephen Sommers, a particle accelerator is operated with the scientists a mere few feet away from the beam, standing behind what looks like a glass partition. While it is certainly true that focusing on the ionizing radiation from a particle

Fig. 3.3 Movie poster for *The Attack of the 50 Foot Woman* (1958) (Public domain)

accelerator grossly misrepresents the machine's main purpose, and converts the average person's fears of radiation into fears of particle accelerators, we should also avoid flagrant departures from the truth in the other direction. A healthy respect for ionizing radiation will certainly go a long way in improving your health!

3.4 Malevolent Magnets

Sean Carroll notes in *The Particle at the End of the Universe*, "Magnets reach out, across apparently empty space, to pull things toward them. Kind of freaky, when you think about it" [14]. Magnets seem almost magical, as anyone who has given magnetic toys to a child can attest. At the same time, the general public willingly spends an estimated $1 billion per year worldwide on magnetic cure-alls that lack any scientific evidence to back up their claims, including magnetic bracelets, shoe inserts, pillows, and even magnetic fuel treatments supposed to increase fuel efficiency [15]. Driving these beliefs may be a common misconception that human blood is magnetic, since it contains iron in its hemoglobin. If this were true, then it would be easy to separate red and white blood cells by using a magnet, which is not the case. If blood were magnetic, MRI, or Magnetic Resonance Imaging, machines would be used as instruments of torture rather than diagnosis [15]. MRI is an imaging technique in which a magnetic field similar in strength to that of the LHC dipole magnets is used to make the protons in the hydrogen atoms in your body's water line up in a particular way. Afterwards, a radiofrequency (RF) electric field is applied (similar to one in an accelerator), which makes some of the protons flip over, producing an electric current that is picked up by the machine and turned into a picture. It's all perfectly safe unless you happen to be wearing metal that can be attracted to the magnet, as demonstrated in very dramatic online videos![1] It is important to note that MRI used to be called NMR, Nuclear Magnetic Resonance, but the terminology was changed because "Nuclear" makes it sound similar to a nuclear bomb, and therefore dangerous.

Depictions of magnets in popular culture are also fraught with mystery and the threat of potential danger. From the film *The Core*, based on the supposed cataclysmic disruption of the Earth's magnetic field, to the arch villain Magneto of *X-Men*, anything dealing with magnets has the potential to be bad news. Perhaps nowhere is this more explicit than in the strange magnetic properties of the unnamed island in the TV series *Lost* (2004–10). From compasses

---

[1] https://www.iflscience.com/health-and-medicine/how-dangerous-are-magnetic-items-near-mri-machine/

pointing in the wrong direction to pockets of deadly electromagnetism pulling airplanes from the sky, *Lost*'s exploitation of magnetism is legion [16].

In contrast, the magnets of the LHC should inspire awe and not apprehension, as they represent a triumph of engineering. This is especially true of the 2500 dipole magnets that keep the accelerating particles turning at the correct radius within the beam tube. At 15 meters long each with an individual weight of 35 metric tons, these are impressive beasts, to say the least. As noted in Chapter 1, these superconducting electromagnets operate at super cold temperatures in order to maximize efficiency, generating an impressive 8.3 Tesla (T) magnetic field [8]. For comparison, the magnetic field of the average sunspot (considered a magnetic "storm" on the sun) is about 0.2 T. Not coincidentally, the magnets from the S.T.A.R. Labs accelerator that are used as door locks to the RF cavities turned prison cells for the meta-humans in *The Flash* have superconducting electromagnets with a field strength of, you guessed it, 8.3 T.

While the LHC dipole magnets do not produce anywhere near the strongest manmade magnetic fields ever achieved, they are central to controlling the LHC's particle beams and are certainly worthy of respect. As CERN discovered soon after the machine's inaugural start-up, when those dipole magnets go wrong, there's hell to pay (see Chapter 5). While magnets might seem mysterious to the nonscientist, they are far more familiar than objects beyond our atmosphere. Cosmic wonders can also quickly lead to cosmic fears, especially when the Internet is involved.

3.5   Cosmophobia: The Universe Is a Scary Place

Some time around 2008 I noticed a significant change in the questions posed by children during public outreach events. Rather than ask what happens to something that falls into a black hole (nothing good, I can assure you), children were concerned with something far more immediate—was the world going to be destroyed in 2012? I also found my college students making reference to a 2012 apocalypse, invoking a potpourri of pseudosciences and misconceptions involving the Mayan calendar, asteroids, solar flares, the Earth's magnetic field (there are those pesky magnets again!) and more. One of the first professional astronomers to go on the offensive in combating these unfounded fears was NASA astrobiologist David Morrison, whose "Ask an Astrobiologist" blog became a defacto 2012 Apocalypse debunking headquarters. From the hypothetical alien spacecraft-posing-as-an-asteroid Nibiru to a supposed alignment of the sun with the galactic center, a proposed catastrophic flipping of our planet on its poles (a common misconception of what it means when our magnetic poles switch polarity) to the mother of all solar outbursts, the general public seemed obsessed with the paranoid notion that the universe is out to get us. As the Gregorian calendar approached December 2012, Morrison, myself, and others, devoted increasing time and effort to not only debunking these pseudosciences and just plain lies, but calming public fears. As Morrison astutely noted, the increasing "cosmophobia"—fear of the cosmos—had the potential to leave a permanent scar on the public psyche [17]. Cosmophobia has been successfully exploited by numerous authors and screenwriters who connect the fear of cosmic objects (mostly connected to the violent deaths of stars) with fears of particle accelerators, a double whammy.

Once a star reaches the end of its energy generating fusion cycles, the exact nature of its corpse depends on its mass. Low mass stars collapse to about the size of the Earth, forming a dense ball of mainly helium, carbon, and oxygen called a *white dwarf*. Further collapse is prevented by a previously described quantum effect, the Pauli exclusion principle. Gravity can cram together the electrons' available quantum states, just as the seats in a classroom can be shoved closer together, but there comes a point where neither can be compactified any further. This halts the gravitational collapse of the star and holds up the white dwarf against further collapse. The strength of this electron degeneracy pressure can only counter gravity if the star dies with a mass under 1.4 times the mass of the sun, named the Chandrasekhar limit after Nobel Prize laureate Subrahmanyan Chandrasekhar who first calculated it. Stars that die with greater masses explode as a *supernova*.

If the core of the dying star remains intact and is lighter than about twice the mass of the sun, the electrons combine with protons to form neutrons, with a wind of neutrinos traveling near the speed of light fleeing from the dying star hours before the explosion is seen at the star's surface. This type of stellar corpse is therefore composed of neutrons, and their degeneracy pressure prevents further collapse once the corpse has compressed to the size of a large city. The first neutron stars were discovered by British graduate student Jocelyn Bell in 1967 as part of a radio survey. As described earlier in this chapter *pulsars* are synchrotron radiation-emitting neutron stars rotating many times a second.

For dying stars heavier than about twice the sun's mass, there is no force in the universe that can halt the collapse, and the material implodes to a mathematical point called a *singularity*, where the laws of physics break down. Such an object's gravitational field is so strong (or in the language of Einstein, it warps space-time to such a high degree) that not even light can escape, forming

Fig. 3.4 An artist's conception of a black hole in a binary system devouring its companion star (ESO/L. Calçada/M. Kornmesser, CC BY 4.0, via Wikimedia Commons)

a black hole. A black hole is not directly observable; instead, we study its influence on its neighborhood. In the case where a black hole and a normal star are in orbit around each other, the black hole pulls gas off its companion, which spirals around as if it were going down a drain. This swirling material forms an *accretion disk* around the black hole. The kinetic energy of these particles is so high that it heats the disk to a million degrees, creating X-rays that astronomers can observe. It is also not surprising that the public, while fascinated by black holes, has serious misconceptions concerning them (for example that they cruise through the universe like some cosmological Megalodon shark looking for prey on which to feed). In response, scientists have posted multiple webpages debunking these misconceptions [18]. Black holes of stellar size (typically less than 20 times the mass of the sun) are certainly fascinating creatures, but truly monstrous black holes weighing the same as a million suns lurk in the hearts of many galaxies (Fig. 3.4).

While the black hole at the center of the Milky Way is currently rather mild-mannered as far as galactic black holes go (reminiscent of a middle-aged house cat that is well taken care of and lazily lounges around the house), in our galaxy's youth the higher density of material close to the central black hole proved to be low hanging fruit—literally. Once upon a time, the accretion disk at our galaxy's core was a powerhouse of electromagnetic radiation. Today we see similar behavior in other galaxies billions of light years away, because the light we see tonight left that galaxy billions of years ago. We are therefore looking into the past of that galaxy, to its earlier toddler tantrums. Before you feel smug, intelligent beings in that same galaxy are probably viewing our galaxy tonight as it was in its "terrible twos" (two billion years old, that is). Galaxies viewed in these earlier energetic stages of existence are generally termed *active galaxies*. A particular subclass of these objects, the quasi-stellar object or quasar, confounded astronomers in the early 1960s when they were first discovered, as only the bright cores of these hyperactive galaxies were originally visible in telescopes. In the *The Outer Limits* episode "Production and Decay of Strange Particles" Dr. Marshall carries out cyclotron experiments in which he bombards nobelium-238 with particles from quasars and creates an ultra-heavy highly radioactive isotope that causes "a hole torn in the universe". Marshall says of quasars "They radiate. They pulse. But, they're not galaxies and they're not stars. They break every rule in the book. Yet they're out there—burning" [19]. It is interesting to note how this science fiction series reflects our incomplete understanding of quasars at the time of its airing. There is no such excuse available for the screenwriter of *The Void*, in which Dr. Thomas Abernathy seeks to generate electrical power from a "miniature controlled quasar" [20]. What he is really speaking about here is just a tiny black hole that he creates in his accelerator (but cannot control).

A miniature black hole is nothing like a miniature poodle. Proposed by physicist Stephen Hawking to have been created in the extreme conditions of the early universe, these so-called Hawking or primordial black holes pack the mass of a mountain into a volume the size of an individual proton. Like a cosmic drill bit, such an object could pierce one side of our planet and travel straight through like a knife through butter. Research suggests that the resulting cylinder of damaged rock would be recognizable in the geological record [21]. But if a black hole were to be created here on Earth, it would not have enough speed to escape the planet's gravity, and instead would continuously dive down through the Earth, emerge about half a world away, then repeat this motion as it grew on a steady diet of rock and metal. Ignoring any friction (which would slow it down), the time required for one complete back and forth is about 90 minutes. As we shall see, numerous works of science fiction incorporate such a munching menace burrowing into our planet. For the moment let's note two examples. In Dan Simmons' 2005 novel *Olympos*, Paris Crater is a city in the fourth millennia built on what was left of Paris after an uncontrolled microscopic black hole created at the Institut de France sank to the core of the Earth and took much of the Louvre with it. If a mini black hole were to be created with some initial downward speed (for example, if it was released from some height above sea level), it would emerge from the Earth on each pass, climb up to its original height (again ignoring friction), and then fall down, puncturing a fresh hole in the planet because the Earth has turned slightly as the black hole was busy doing aerial acrobatics. For example, the black hole created by scientist Paul Krone in J. Craig Wheeler's 1986 novel *The Krone Experiment,* said to "eat and grow like a cancer in the bowels of the earth", is made even more terrifying in the film version when it threatens to pass down directly through the San Andreas Fault [22]. Scientists decide it is better to take the risk to shoot a Russian space laser at the black hole as it passes over nearby Imperial Valley in order to shift its course just slightly, resulting in an earthquake of "only" 7.2 on the Richter scale.

But the behavior of tiny microscopic black holes is more complicated than this. In the 1970s Stephen Hawking pondered the effect of quantum mechanics on the behavior of black holes. To his surprise, he found that, unlike their big brothers, any miniature menaces found in our universe today should naturally get smaller, not larger, radiating away their mass at an ever-increasing rate until they ultimately explode in a shower of gamma rays with the equivalent energy of roughly a million megatons of TNT. Gamma ray astronomers could, theoretically, observe this. It should be emphatically stated that there have, as of now, been no observations of either microscopic black holes or the so-called *Hawking radiation* that this Hawking mechanism of black hole evaporation predicts. However, as we shall return to at several points in our story, physicists expect that Hawking was right about all this.

While the Hawking mechanism allows for a way to get rid of pesky primordial black holes (and those we might accidentally or intentionally create in a particle accelerator), the resulting shower of ionizing radiation doesn't exactly sound like a good time. In his DVD commentary to the 2005 film version of his novel *The Krone Experiment* (directed by his son, J. Robinson Wheeler), physicist J. Craig Wheeler admits "There's some other things I've left out, that I don't talk about too much, like Hawking radiation, we kinda ignored. We allude to it, but in fact it could be very dangerous, so we kinda left that outta the book plot" [23]. However, in the 2012 sequel, *Krone Ascending*, the radiation associated with black holes plays a central (and deadly) role. In the opening vignette of Thomas Wren's *The Doomsday Effect* (1986), the passage of a microscopic black hole through an airplane leaves two passengers suffering from radiation sickness. On another passage up through the Earth, the black hole irradiates a hog farm, giving at least one of the poor animals a lethal dosage. The take away message is that a black hole can create cancer-causing X-rays and gamma rays, either though the accretion disk generated by material spiraling into it as it eats, or through its evaporation thanks to Hawking radiation. Either way, you don't want to be anywhere near a black hole (Fig. 3.5).

Fig. 3.5 Chandra X-ray satellite image of the X-ray emissions from the famous black hole Cygnus X-1 (NASA/CXC, public domain)

Another source of cosmophobia is also hypothetical at present, namely the intervention of belligerent extraterrestrials. Pop culture representations of aliens seem more often aligned with *Starship Troopers, Alien,* and *The Predator* than *Lilo and Stitch, ET,* and *Contact.* Even *Doctor Who,* the longest running science fiction television series in the world, portrays many extraterrestrials as both unhuman and inhumane (e.g. the Daleks, Jagrafess, and Slitheen). Three studies done by researchers at Arizona State University suggest that the discovery of microbial ETs would be positively received by the majority of Americans, but would the same be true of the discovery of potential technological rivals (or superiors) [24]? No less than Stephen Hawking famously stated that any space traveling species that could reach us would far exceed our technological prowess, and would probably treat us in much the same way that humans of the past treated technologically less advanced groups of our own species—in other words, very badly [25]. Others have argued that any species that has reached such an advanced stage of technology must have overcome their baser nature, or else they would have destroyed themselves already through nuclear war or its extraterrestrial counterpart.

Fears of world domination or destruction by extraterrestrials have led some to question the wisdom of Messaging to Extraterrestrial Intelligence (METI), currently done through intentionally sending out radio signals and placing plaques and other messages on the sides of our interplanetary space probes. One of the complaints lodged by opponents is that METI projects do not have to go through peer review in order to judge whether the benefits far outweigh the risks (as in the case of medical experiments) [26]. These same charges have been leveled against particle physicists when higher and higher energy particle colliders have been built. But others have argued that not only is hiding from technologically advanced ETs impossible, but we run the opposite risk of missing out on beneficial knowledge and technology that we could receive from an advanced species that has outgrown its violent adolescence [27]. Perhaps it is we who will be judged unworthy of admission into the universal family, as the fictional citizens of South Park, Colorado learn the hard way when Stan Marsh illegally uses a piece of an LHC dipole magnet to cheat on his son's pinewood derby car. When the soup-up vehicle is unintentionally launched into space, the extraterrestrial welcome wagon tests humanity and we receive a failing grade, leading to Earth being quarantined to prevent our infecting the rest of the universe with stupidity and greed [28].

The concept of a particle accelerator experiment bringing our planet to the attention of extraterrestrials is front and center in John Ringo's *Into the Looking Glass* and John Cramer's *Einstein's Bridge*. In both cases it is the successful creation of Higgs bosons that catches the attention of alien invaders and benevolent ET allies who come to our aid in our time of need. But no such aid comes to *The Outer Limit's* Dr. Marshall after his cyclotron experiment with particles from quasars takes an ugly turn and invites harmful energy beings into our dimension. Yet again we see how in popular culture common anxieties + particle accelerators = trouble for humanity.

3.6   Case Study: The Tunguska Incident

Natural disasters are another relatively common trope in science fiction media. From tornadoes and earthquakes to asteroid collisions and volcanic eruptions, Hollywood has capitalized on them all. Not satisfied to merely scare the public with run of the mill catastrophes, popular culture has dreamed up exotic mash-ups of disasters, such as *Atomic Twister* (tornadoes causing a meltdown at a nuclear power plant), *Lavalantula* (lava plus killer tarantulas), and the infamous *Sharknado* series of shark infested tornados. A real-world event that occurred in Siberia in 1908 is frequently paired with particle accelerator disasters in film and novels, perhaps because of the air of mystery that still surrounds it in some quarters.

In director Ivan Reitman's 1984 film *Ghostbusters* Dr. Ray Stanz explains to Louis Tully after the closing of the door to Gozer the Gozerian's dimension "You have been a participant in the biggest interdimensional cross rip since the Tunguska blast of 1909" [29]. The so-called Tunguska incident actually occurred on June 30, 1908 in a remote pine forest in the Podkamennaya Tunguska River area of Siberia. Something caused an explosion with the energy of a large hydrogen bomb that utterly destroyed nearly a thousand square miles of trees. In 1930 noted comet specialist Fred Whipple and Russian meteor researcher Igor Astapovich independently proposed that an astronomical object, either a comet or large stony meteoroid, was responsible for the event. The lack of a crater argues against a denser iron meteoroid, which would have had a greater chance of surviving to hit the ground. While a comet or meteoroid is still the most widely accepted explanation among scientists today, a number of more speculative suggestions have been made, many of which combine Tunguska with particle physics [30]. These include a large amount of antimatter (an "anti-rock") exploding upon contact with our atmosphere [31], a ball of dark matter causing an explosion within the Earth [32], and a Hawking black hole [33]. There have even been suggestions that an extraterrestrial spacecraft self-destructed or crashed [34] and that the eccentric scientist Nikola Tesla's experiments were somehow involved [35] (Fig. 3.6).

Speculations about the Tunguska event often appear in novels and films in which a particle accelerator plays the role of the villain. For example, the novel *God's Spark* opens with a flashback revealing that the Tunguska incident was caused by a rogue electromagnetic experiment by Tesla's disgruntled assistant half a world away in the mountains of Colorado. When the Russians steal the black hole/anti-black hole particles from CERN and bring them back to Siberia there is another Tunguska-like explosion. Frequently in fiction when miniature black holes are created by a particle accelerator a comparison is drawn to Tunguska and the devastation caused in that event. In other words, if *that* black hole could do so much damage, think how much *our* black hole will do! Conversations to this effect are found in film *The Void* and the novels *The Krone Experiment* and David Brin's *Earth* (2005). In this last work, it gets even more complicated. While Dr. Alex Lustig creates a black hole dubbed Alpha that becomes lodged within the Earth, it is not the true threat. That

Fig. 3.6 Trees devastated by the Tunguska incident (Public domain)

honor belongs to Beta, a larger and rapidly growing black hole that is found to have first entered the Earth in Siberia in 1908. The scientists realize to their horror that Beta caused the Tunguska incident, and is an alien-created singularity that poses a major threat to humanity. Alex reasons that it was sent to Earth only a few decades after the first human radio signals were sent into space, and the extraterrestrial receivers of our unintentional message might not be coming in peace. In this particular case, the opponents of METI would say "See—I told you!".

The apparent lesson of several of these works therefore aligns with the opinions of the opponents of intentional communication with extraterrestrials. But we can draw broader lessons as well. Whenever there is communication, there is the possibility of miscommunication. This is certainly true when scientists communicate with the general public. Having noted several sources of anxieties that can complicate such communication, we will now expand the orbit of our survey to include the broader issue of science communication and how to avoid the most common pitfalls.

References

1. M. Caiden, *Star Bright* (Bantam Books, New York, 1980), p. 138
2. A. Helbing, T. Helbing (script), The Race of His Life, The Flash, season 2 (2016)

3. D. Wojcik, Embracing Doomsday: Faith, Fatalism and Apocalyptic Beliefs in the Nuclear Age. West. Folk. **55**(4), 321 (1996)
4. Jesus Christ's Return to Earth, Pew Research Center, http://www.pewresearch.org/daily-number/jesus-christs-return-to-earth/
5. C. Raasch, For 'Preppers', Every Day Could Be Doomsday, USA Today, http://www.usatoday.com/story/news/nation/2012/11/12/for-preppers-every-day-could-be-doomsday/1701151/
6. G. Edwards (dir.), End:Day, BBC (2005)
7. A.D. Aczel, *Present at Creation* (Crown Publishers, New York, 2010), p. 9
8. FAQ: LHC the Guide, CERN, https://cds.cern.ch/record/2255762/files/CERN-Brochure-2017-002-Eng.pdf
9. A. Strickland, 'Ghost Particle' Found in Antarctica Provides Astronomy Breakthrough, CNN, https://www.cnn.com/2018/07/12/world/neutrino-blazar-cosmic-ray-discovery/index.html
10. N. Scharping, What Keeps an Astronaut Awake at Night? Cosmic Rays, Discover, http://blogs.discovermagazine.com/d-brief/2017/12/19/astronauts-cosmic-rays
11. G. Benford, *Cosm* (Orbit, London, 1998), p. 17
12. J.B. Lowder, Why a Zombie Movie Made by Physicists Is the Best Kind of Science PR, Slate, http://www.slate.com/blogs/future_tense/2012/12/12/decay_a_zombie_movie_created_by_scientists_and_filmed_at_cern_video.html
13. S. Neumann, Three Misconceptions About Radiation: And What We Teachers Can Do to Confront Them. Phys. Teach. **52**, 357 (2014)
14. S. Carroll, *The Particle at the End of the Universe* (Plume, New York, 2012), p. 116
15. L. Finegold, B.L. Flamm, Magnetic Therapy. Br. Med. J. **332**, 4 (2006)
16. E. McCarthy, Lost Abuses the Laws of Electromagnetism, Popular Mechanics, https://www.popularmechanics.com/culture/tv/a5647/lost-happily-ever-after-fact-check/
17. D. Morrison, Doomsday 2012, the Planet Nibiru, and Cosmophobia, Astronomy Beat, no. 32 (2009), pp. 5–6
18. A. Bauer, C.A. Onken, Black Hole Truths, Myths and Mysteries, Australian Academy of Science, https://www.science.org.au/curious/space-time/black-holes; The Truth and Lies about Black Holes, Chandra X-ray Observatory, http://chandra.harvard.edu/resources/podcasts/ts/ts301107.html
19. L. Stevens (script), Production and Decay of Strange Particles, The Outer Limits, season 1 (1964)
20. G. Shilton (dir.), The Void, Lions Gate Entertainment (2001)
21. I.B. Khriplovich, A.A. Pomeransky, N. Produit, G. Yu. Ruban, Can One Detect Passage of a Small Black Hole Through the Earth?, ArXiv, https://arxiv.org/abs/0710.3438
22. J. Craig Wheeler, *The Krone Experiment* (Grafton Books, London, 1986), p. 269
23. J. Craig Wheeler, Commentary, The Krone Experiment: Widescreen Director's Cut, DVD, J. Robinson Wheeler (2005)

24. J.Y. Kwon, H.L. Bercovici, K. Cunningham, M.E.W. Varnum, How Will We React to the Discovery of Extraterrestrial Life? Front. Psychol. **8**(art. 2308), 1 (2018)

25. R. Boyle, Aliens Exist, and We Should Avoid Them at All Costs, Says Stephen Hawking, Popular Science, https://www.popsci.com/science/article/2010-04/hawking-aliens-are-out-there-and-want-our-resources

26. J. Gertz, Reviewing METI: A Critical Analysis of the Arguments, ArXiv, https://arXiv.org/abs/1605.05663

27. D.A. Vakoch, In Defense of METI. Nat. Phys. **12**, 890 (2016)

28. T. Parker (script), Pinewood Derby, South Park, season 13 (2009)

29. I. Reitman (dir.), Ghostbusters, Columbia Pictures (1984)

30. B. Napier, D. Asher, The Tunguska Impact Event and Beyond. Astron. Geophys. **50**(1), 18–26 (2009)

31. C. Cowan, C.R. Alturi, W.F. Libby, Possible Anti-Matter Content of the Tunguska Meteor of 1908. Nature **206**, 862 (1965)

32. C.D. Froggatt, H.B. Nielsen, Tunguska Dark Matter Ball, ArXiv, https://arXiv.org/abs/1403.7177v3

33. A.A. Jackson IV, M.P. Ryan Jr., Was the Tungus Event Due to a Black Hole? Nature **245**, 88–89 (1973)

34. J. Oberg, Russians Add New Twist to Old UFO Myth, NBC News, http://www.nbcnews.com/id/5686713/ns/technology_and_science-space/t/russians-add-new-twist-old-ufo-myth/

35. Mysterious Tunguska Explosion of 1908 in Siberia May Be Linked to Tesla's Experiments of Wireless Transmission, Tesla Memorial Society of New York, http://www.teslasociety.com/tunguska.htm

4

Can You Hear Me Now? Impediments to Science Communication

4.1 Debunking 101

Imagine, if you can, that workers at CERN have begun talking to dead people, thanks to the LHC. This is the plot of Franklin Clermont's novel *The Voices at CERN*. Shannon Fields, the fictional CERN senior press relations coordinator, is tasked with speaking to the concerned wife of one of the technicians. Shannon's superiors reason that she is qualified for this task because her job entails explaining "each new discovery to the waiting news organizations that felt compelled to deliver news in 30 second sound bites only and language pegged at an eighth grade reading level or lower" [1] as well as talking to the general public about black hole fears and "that destroy-the-world stuff" [2]. Shannon is understandably concerned about the ramifications to their public image (especially on safety issues) if the employees' claims becomes public, imagining the resulting "lurid American tabloid headlines: 'CERN: I Hear Dead People'" [3]. While this example is clearly outlandish, the fictional headline is an excellent parody of reality. However, studies suggest that it is not the tabloids that Shannon and her real-world counterpart have to be concerned about, but rather the Internet, especially in the United States [4]. The fact-checking website Snopes.com has had to debunk several curious Internet claims about CERN, perhaps the most bizarre being a video purporting to show a ritual human sacrifice conducted on the grounds of the facility [5]. In reality it was a prank conducted without the knowledge or permission of CERN and those responsible were held accountable [6]. However, the damage has already been done, as the video is still easily found online.

© Springer Nature Switzerland AG 2019
K. Larsen, *Particle Panic!*, Science and Fiction,
https://doi.org/10.1007/978-3-030-12206-5_4

The seriousness of this rather extreme example becomes apparent when you realize that, while in 2001 only nine percent of Americans polled relied on the Internet as their primary source of news on science and technology, this percentage mushroomed to 55% in 2016 [4]. Given recent concerns over fake social media accounts being used to promulgate equally false political information, the old adage "trust but verify" has taken on greater significance in the case of any information found online. When scientists come across scientific misrepresentations or misconceptions their initial instinct is to offer a correction to the error, often rather vehemently. After all, science is a self-correcting endeavor based on the weight of evidence. However, effective communication with nonscientists is more complicated than arguing about the interpretation of data with your colleagues. As John Cook and Stephan Lewandowsky explain in *The Debunking Handbook*, unless care is taken, debunking can have the opposite of the intended effect, and actually reinforce rather than dislodge these misconceptions.[1]

The most common error made by well-meaning scientists and other trying to correct erroneous views of science is the assumption that belief in misconceptions is simply due to a lack of knowledge on the part of the individual—they simply don't know any better. If this is the case, then just giving someone correct information—and the more information, the better—should solve the problem, right? Unfortunately not. This so-called "information deficit model" does not take into account the way that human beings process information, the central role that personal beliefs play in such processing, and the uncomfortable reality that once miscomputations are formed, they are exceedingly hard to dislodge [7].

Well-intentioned debunking most often backfires due to one of three overreactions on the part of the debunker. The "Familiarity Backfire Effect" occurs when the debunker mentions the misconception so frequently that it becomes more familiar than the truth. To combat this, concentrate on the fact, not the fiction. Related to this is the "Overkill Backfire Effect", in which you provide too many reasons or counterarguments and overwhelm the person whose views you are trying to correct. This is definitely a case where less is more [7]. It is extremely important that our explanations are clear, straightforward, and avoid denigrating personal worldviews that are not central to the argument (such as religious views). Ad hominem attacks (e.g. referring to people as "crackpots") are counterproductive and should be avoided at all cost, although are far too common. For example, Shannon Fields of *The Voices at CERN*

---

[1] This excellent source should be on the desktop (literal or virtual) of anyone engaging in science communication.

muses to herself that her job includes dealing with "the occasional crackpot or 'end of the world' lunatic" [1]. This last point is of the utmost importance, and we will return to it several times throughout the remainder of this book, as it feeds into the "Worldview Backfire Effect", what Cook and Lewandowsky warn is the most important of the three [8]. This occurs when the debunker provides information that openly contradicts the beliefs of the listener. It is strongest in fervent believers, and we just have to accept that some people cannot have their minds changed by the facts. Instead, our time is best spent trying to reach those who are undecided or only minimally invested in a particular point of view. Not only is this a more effective use of our time and efforts, but avoids reinforcing suspicions and ill will.

It is also sometimes possible to open a person's mind by framing the argument in less threatening or less absolute terms. Cook and Lewandowsky use the example of proposing a "carbon offset" versus a "carbon tax" [8]. Framing can be used to minimize or denigrate the opposing position, for example by using terms such as "frankenfood" to describe genetically modified organisms or labelling oil companies as "big oil" [9]. Caution needs to be used when adopting framing as a strategy, because it can backfire. Recall in Chapter 3 that the phrase "mini Bang" or "mini Big Bang" is colloquially used to describe experiments at particle colliders. While the term is certainly catchy, it carries with it considerable baggage (the ability to reinforce fears that scientists are trying to recreate the actual Big Bang) and perhaps should be avoided. Note that the CERN FAQ directly addresses such fears, which suggests that they are rather common [10]. A number of other framing phrases are often used by opponents of the LHC, such as "Earth-killing black hole", "doomsday machine", and "killer strangelet". The anxiety causing "Higgs abyss" was strangely first seen in an editorial on the American Physical Society website describing a new study that actually made a potential doomsday scenario *less* likely rather than more so [11] (Fig. 4.1).

In summary, Cook and Lewandowsky recommend the following four points of emphasis for effective debunking: concentrate on the key facts while avoiding the aforementioned backfire effects; flag any mention of the misconceptions/falsehoods with a clear, explicit warning; clearly and succinctly explain why the misconception is wrong as well as the motives of those who promote it; and utilize visual aids such as infographics. This last point is especially important when numbers are involved. Throughout the rest of this work we will refer back to these pitfalls as examples warrant. For the moment, the remainder of this chapter further explores several of these issues in depth, utilizing examples from both fictional and real world particle physics (Fig. 4.2).

Fig. 4.1 The phrase "big oil" is used as a framing phrase by protestors against tax breaks for oil companies (AFL-CIO, CC BY 2.0, via Wikimedia Commons)

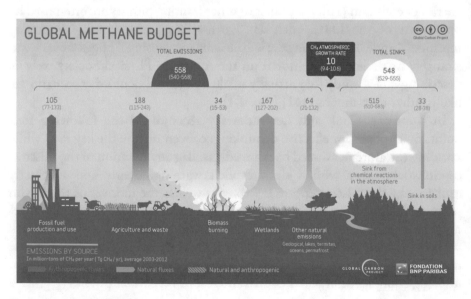

Fig. 4.2 An infographics representation of the world's methane budget (Global Carbon Project, CC BY-SA 4.0, via Wikimedia Commons)

4.2   Science Literacy and Numeracy

Detective Joe West often provides comic relief as the avatar for the audience on *The Flash* whenever the show's scientists engage in technical explanations. For example, when forensic scientist Barry Allen attempts to explain how he created another copy of himself using his incredible speed, Joe's son Wally has an expression of absolute bewilderment painted across his face. His father questions "Is that what I look like when they start talking about science?" [12]. Darien Fawkes, the titular character of the 2000–2002 television series *The Invisible Man*, is a special agent for the Bureau of Weights and Measures, and turns to a fictional book "Particle Physics for Imbeciles" (an obvious take off on the famous "For Dummies" series) for basic background information when he investigates a crime involving a particle accelerator. He follows this up by asking thoughtful questions based on his reading to scientist Dr. Claire Keeply [13]. In my personal experience, any scientist is far more likely to come across a Wally West than a Darien Fawkes when discussing science with friends, family, or the general public. This view is bolstered by a 2014 Pew Research Center study of nearly 4000 members of the American Association for the Advancement of Science. The study found that 84% of AAAS scientists thought that the public's limited knowledge of science was a major problem, with 14% calling it a minor problem. When asked for a probable cause for this 75% thought that a major reason was that there is not enough STEM in K-12 education, with 22% calling this a minor reason. Interestingly, 40% thought that a major reason was that not enough scientists are communicating their findings, with 49% calling this a minor reason [14].

Just how limited is the American public's understanding of science? The National Science Board has been publishing the results of national surveys of science literacy for decades, as part of their *Science and Engineering Indicators* volumes. Part of this data comes from a short true/false questionnaire of science facts, two questions being especially relevant to our discussion of particle physics: "All radioactivity is man-made", and "Electrons are smaller than atoms". In the most recent study (2016), 70% and 48% of Americans surveyed correctly answered that these statements are, respectively, false and true. The average number of questions correctly answered out of nine has remained stagnant (between 5.6 and 5.8 or 62 and 64%) since 2001 [15]. Having the American public stuck at an average grade of "D" is discouraging, especially given the important social and political issues that are deeply scientific in nature (such as climate change and energy policies). While knowing more science does not necessarily make an individual more supportive of scientific points of view due to the strong influence of personal values and beliefs such as religion, possessing a greater understanding of science is certainly helpful when considering highly technical issues. Therefore having a scientifically literate society is beneficial for society at large.

Science literacy is not solely defined by knowing specific facts, but also includes an understanding of how scientists "work" and an ability to thoughtfully evaluate how scientific advances (and their technological spinoffs) affect society. Being able to weigh the risks of scientific experiments such as those associated with particle physics and particle accelerators involves more than merely a basic comprehension of the science involved, but also understanding the statistics that form the basis of risk assessments. Mathematics is the language of science, including issues of error and uncertainty that are associated with all scientific experiments. It is not merely science literacy that concerns us here, but mathematical literacy, or what is termed *numeracy*. A 2012 study of the average numeracy of American adults found that only 48% scored at a Level 3 (out of 5) or higher [16]. Among the Level 3 skills are "interpretation and basic analysis of data and statistics in texts, tables, and graphs" [17]. From the misuse of crime statistics to misinterpretations of medical risks, the apparent inability of the average American to properly interpret statistics and probabilities is a serious problem. This also has serious repercussions for communicating the results of safety studies of proposed experiments such as those at the LHC and other particle colliders. For example, what scientists would expect to be the most natural way to present this data (the "texts, tables, and graphs" listed above) is actually the least effective methodology when trying to explain the safety of an experiment to the general public. Research suggests that instead it is the accompanying storyline (the explanation in words) that most greatly influences those with lower numeracy skills [18]. We will return to these ideas in future chapters when we review attempts by individual scientists and facilities such as CERN to calm public fears of their high energy experiments through safety studies. The bottom line is that words (and visual aids) matter far more than numbers when trying to effectively communicate with the average nonscientist. The downside of this is that sometimes the words scientists use are just as bewildering as the numbers, as poor Wally West discovered. Nowhere is this more true than the world of quantum mechanics.

4.3   Quantum Weirdness, Pseudoscience, and Fringe Science

In an early scene from John Carpenter's 1987 classic horror film *Prince of Darkness* physics graduate students Walter and Catherine discuss the difficulty of understanding quantum mechanics. Walter complains that quantum mechanics doesn't make sense because it "violates common sense". Catherine not only agrees, but adds that this "is the entire complete point. It doesn't make any common sense. Our common sense breaks down on a subatomic level" [19]. All advanced physics students pass through a similar realization at some point in their schooling. Recall from Chapter 1 that quantum mechanics predicts the existence of antimatter and the borrowing of energy from the vacuum to create virtual particle-antiparticle pairs. As strange as this process may seem, it is perhaps among the most straightforward predictions of the discipline.

The quantum view of microscopic systems such as atoms and subatomic particles replaces the comfortable, deterministic Newtonian view of reality with an interconnected web of possibilities and probabilities. Perhaps the most disconcerting feature of quantum mechanics is the realization that any system is not in an unambiguous state of existence until it is observed. Take, for instance, an experiment where an electron is presented with two equal sized holes in a wall and passes through one to the other side. Before the result of the experiment is observed (before we know which hole the electron ultimately went through) its so-called *wave function* is described as a mixed state of being where it can be said to have gone through *both* holes. In the standard or Copenhagen interpretation of quantum mechanics the observation "collapses" the wave function, a process by which the unrealized possibility disappears. In other words, after observing the electron we know it went through hole B and the part of the wave function describing it as having gone through hole A disappears. In the more radical Many-worlds Interpretation (MWI), both possibilities become "real". There is a copy of the observer and electron in which s/he observes it to have gone through the first hole, and another, just as "real" set of observer and electron in which the electron traveled through the other hole.

In a game of golf there is never any discussion of the ball being in a mixed state of in the hole/not in the hole simultaneously until we actually observe it. That is because when it is scaled up to our everyday level, quantum mechanics suggests nonsensical results. It seems like something out of a science fiction novel. It is exactly this spookiness of quantum mechanics that drove John

Carpenter to write the screenplay for *Prince of Darkness* [20]. In the words of Professor Farnsworth of the adult animated series *Futurama*, "quantum physics means anything can happen at any time for no reason", allowing screenwriters and novelists to invent futuristic technologies or scientific advances simply by the addition of the word *quantum* to the title [21]. Examples include the quantum torpedoes of the *Star Trek* universe and the quantum crystalline armor of the Jedi Academy series of *Star Wars* spinoff novels.

Perhaps the spookiest of all the predictions of quantum mechanics is the potential existence of so-called entangled states between two systems. If we measure some property of a system (e.g. spin), the other system appears to instantly change its property in response, even when they are so far away that a signal traveling at the speed of light could not share this information fast enough. Popular-level explanations usually include Einstein's nickname for it, "spooky action at a distance" [22]. Potential applications of entangled photons include quantum cryptography (creating very secure encryptions of information) and quantum computing [23]. Quantum entanglement sounds both cool and curious at the same time, leading to its incorporation into popular culture. For example, Vittoria Vetra of the novel *Angels & Demons* is a bio-entanglement specialist, a field that appears to be clearly science fiction. However, quantum effects such as entanglement have either been observed or theorized to play a role in photosynthesis in plants, the ability of some bird species to navigate using terrestrial magnetism, and some aspects of vision [24] (Fig. 4.3).

An entanglement of a slightly more down to Earth nature is featured in *The Big Bang* (2010), a film filled with one clever physics reference after another. These include a character named Adam Nova and locations such as Minkowski's strip club, Schrödinger's warehouse (where an adult film called *The Black Hole* is being filmed), and the Planck's Constant Café. The title references not only the beginning of the universe (which the independently wealthy, and clearly mad, scientist/entrepreneur Simon Kestral seeks to explore in his personal eight-mile-long accelerator) but also the explosion of said accelerator when it predictably fails miserably in its attempt to find the Higgs particle. In a rather unconventional sex scene with private detective Ned Cruz, waitress Fay narrates their breathless encounter with a lecture on quantum mechanics and the Standard Model, including quantum entanglement, concluding "It's real-life magic" [25].

Quantum entanglement is also used by the meta-human Shawna Baez in *The Flash* to teleport from one location to another, while quantum bio-entanglement involving the manipulation of time instead of space appears in the series *FlashForward*. Special rings called QEDs—quantum entanglement

EINSTEIN ATTACKS QUANTUM THEORY

Scientist and Two Colleagues Find It Is Not 'Complete' Even Though 'Correct.'

SEE FULLER ONE POSSIBLE

Believe a Whole Description of 'the Physical Reality' Can Be Provided Eventually.

Fig. 4.3 *New York Times* headline from 1935 reporting on the famous paper by Einstein, Podolsky, and Rosen that described "spooky action at a distance" (*New York Times*, public domain, via Wikimedia Commons)

devices—allow the wearer to remain awake during any flashforward blackout event by tethering the wearer's consciousness to the present. The rings bear the Greek letter alpha, which in physics is used to denote the fine structure constant, 1/137. As described in Chapter 1, this is the ratio of the strength of the electromagnetic force to the strong force. While the television series ended before many of its mysteries could be fully explored, such scientific references had been noted by viewers from the beginning [26].

Highly speculative ideas involving quantum mechanics and human consciousness have been suggested by real world physicists, including appeals to higher spatial dimensions to explain paranormal phenomenon such as ESP. While done by serious mathematicians and physicists such as Bernard Carr [27], who studied under Stephen Hawking, and physics Nobel laureate Brian Josephson [28], these studies are considered outside the mainstream of science [29]. Nonscientists have felt free to follow their lead, publishing their own ideas for combining aspects of theoretical physics with the paranormal.

For example Randy Ruyle offers that the extra dimensions predicted by string theory give "a physical explanation for the paranormal events in the movie *The Sixth Sense*" [30]. Such ideas have also made their way into science fiction, where they can be freely explored without being forced into fidelity with the laws of physics.

A case in point is *The Sparticle Mystery* (2011–5), a British science fiction television series that followed the adventures of a group of children after a particle accelerator accident sends all those aged 15 and older into another reality. Cat, the daughter of physicist Dr. Henry Barker, explains that the project was built in a secret location and designed to search for parallel dimensions. One of the children, Reese, is still able to see some of the parents as ghosts, and her psychic powers play an increasingly important role in the series. Dora Petty, a colleague of Barker's, is the lone voice who tried to warn her colleagues about the outcome of their experiment. So-called Doomsday Dora leaves a video for the children instructing them to attempt to realign the two realities. Her explanation of the accident melds both SUSY (supersymmetry, described in Chapter 1) and pseudoscience/spirituality; she offers that every particle has a supersymmetric partner or sparticle, allowing exact copies of everyday objects to exist simultaneously in parallel universes. However, Dora claims that this does not follow for humans, as our heart chakras (one of the seven bodily centers of energy in Eastern traditions) prevent us from existing in multiple universes simultaneously. As the children search for the secret underground location of the accelerator, Reese discovers more "sensitives" like herself who possess powers of telepathy, telekinesis, and the ability to send their astral bodies into other dimensions (possibly echoing some of the claims of Brian Josephson). At the end of the first season, the children fire up the Sparticle particle collider (with Reese using her psychic powers to try and hold together a sabotaged magnet) and the adults materialize in an unstable state. In order to save Reese's life, the experiment is aborted.

In the second season the children follow clues to the location of another particle accelerator, one built by Quantum Nexus Limited to solve the world's energy problems rather than for scientific exploration. Note the use of the word "quantum" here. The parents materialize as teenage versions of themselves, setting the stage for the final season, in which the teen parents and their children fight for control of the world. *The Sparticle Project's* transition from referencing science to more pseudoscience/spirituality is reminiscent of *Lost*, especially in its highly spiritual conclusion, involving a prophesized comet, a lighthouse (perhaps a direct nod to the final season of *Lost*), psychic powers, and a strange old man holding a newborn baby. While *The Sparticle Particle Project,* like *Ghostbusters,* does not pretend to be hard science fiction, the seamless pairing of science with the paranormal blurs the line between science and pseudoscience, something that is especially worrisome in a children's series (since the intended audience is less well equipped to tell the difference between the two). A number of studies have found a relationship between paranormal beliefs and viewing programs with positive portrayals of the paranormal (such as *The X-Files* and *Touched by an Angel*), even if the viewer had not personally had previous paranormal experiences [31]. Since pseudoscientific beliefs are among those misconceptions that are particularly hard to dislodge (such as fervent belief in astrology), such mash-ups can add to already existing difficulties in effectively communicating science.

Another common difficulty arises when a physicist comes to the attention of the general public, perhaps through a news story or a public talk. Too often said physicist will begin receiving unsolicited emails and manuscripts pedaling "alternative theories" to those of Einstein. Physicist and science fiction author Gregory Benford describes fictionalized accounts of such interactions in several of his science fiction novels, scenes that definitely strike a chord with any physicist who has suffered through such encounters. One of the hallmarks of these eye-roll-invoking "theories" is the random usage of actual scientific terminology, equations that follow no known rules of mathematics, and the creation of new terminology, such as "megaworlds" in *Cosm* [32] and "super-ons" in *Timescape* (1980) [33].

On the other hand, true scientific jargon is often a stumbling block to effective communication with the general public. The titles of three papers located at random on the arXiv database—"Axial field induced chiral channels in an acoustic Weyl system", "Supersymmetric polarization anomaly in photonic discrete-time quantum walks", "Breathing mode frequency of a strongly interacting Fermi gas across the 2D–3D dimensional crossover"—sound suspiciously like *scientific terminology salad*, and it is difficult for the general public to tell the difference between these bona fide scientific papers and pseudoscience meant to bamboozle them.[2]

Countless screenwriters and authors have invented their own science-sounding terminology, not only by simply adding the word "quantum", as previously discussed, but by brute force. For example, in *Xtro II* a particle accelerator generates a nonsensical "neutrino-rich proton beam that can travel the speed of light" as part of the process of firing up the "interdimensional bi-location core dual tangents" [34]. In the direct-to-video film *Futurama: Bender's Game* (directed by Dwayne Carey-Hill) Professor Farnsworth explains

---

[2] The numerous hoaxed papers published in peer-reviewed humanities journals demonstrate that the problem of distinguishing jargon from gibberish is not unique to science.

how he was colliding simple dark matter in a particle accelerator "in an ill-conceived attempt to create a more durable harpsichord wax", and "against all probabilities" a large explosion created both a "non-local metaparticle crystal" that links together all dark matter in the universe (now in a new crystalline form), and "an opposite crystal made of pure anti-backwards energy" [21].

A particularly extreme example is a work that openly admits its use of scientific terminology salad, James M.M. Baldwin's 2012 online short story *The Collider*. Baldwin's story also deserves an award for its ability to draw upon so many of the common physics fears referenced in Chapter 3. Here CERN physicist Daniel Bradford saves the world using the 8th grade science project of Sage, the teenage son of the CERN Chief of Operations. This particle accelerator/transporter built out of copper tubing, wires, plywood, and pieces from high-definition three-D movie players, comes in handy when the attempt to generate a quark-gluon plasma at CERN instead creates an uncontrollable stream of fictional "Magios, the smallest known particles" that is directed into the supermassive black hole at the center of the Milky Way [35]. Bradford explains to a reporter that a "sub-atomic collision designed to fragment the coherence of constituent particles has cascaded into a successive reaction, which, in turn, is dissipating the stability of our planet's mass", to which the reporter demands a jargon-free explanation for the general public. Bradford simply offers, "If we don't find a way to reverse it, the planet will become uninhabitable in forty-eight hours and will cease to exist within a week" [35]. Baldwin pretty much hits the nail on the head here; much of science communication is either inappropriately technical or so grossly oversimplified to the point of bordering on being wrong. Popularizers of science need to be aware of how difficult it is to thread the needle, so to speak, and avoid both pitfalls.

There is also another potential problem here. The ease with which real quantum effects can be coopted into fictional applications, and realistic sounding fictional jargon invented, could potentially confuse the general public, not into mistaking a science fiction novel for a particle accelerator lab manual, but rather when dealing with claims of experimental dangers related to real world accelerators on the Internet and in other media. This could easily lead to baseless fears and mistrust of both the scientists and their experiments, as discussed in the next chapter. As we have already explained, once a misconception has taken root in an individual's brain, it is exceedingly difficult to dislodge.

The blame does not solely rest with those who write popular media, but rather with scientists as well, especially those who communicate with the general public. While scientists are guilty of the sloppy usage of a number of important terms, such as theory when they mean a working hypothesis or model, one of the most dangerous is the term "believe" because it confuses having overwhelming scientific evidence to support a claim with having faith in it [36]. As we shall explore next, there are enough problems between science and religion as it is; why muddy it further with ambiguous terminology?

4.4   Beliefs, Part 1: Religion

The relationship between science and religion has been problematical, at best, since at least the time of Copernicus, a factor we should be extremely sensitive to even today. Since personal beliefs can complicate the dislodging of erroneous information even in the face of scientific facts, it is important to avoid cementing misconceptions by insulting or deriding a person's faith in the name of debunking. Stephen Jay Gould advocated for a relationship of "respectful noninterference" through what he called the Principle of NOMA, or Non-Overlapping Magisteria, whereby the empirical realm of the scientific method does not overlap with the faith-based realm of religion. As Gould succinctly put it, "science studies how the heavens go, religion how to go to heaven" [37]. Any individual is free to ignore the findings of science; however, when one demands that the teaching of religious beliefs is equivalent to established scientific fact (such as claims for a 6000-year age for the Earth), conflicts will necessarily arise. On the other side, claims from scientists that the scientific method, rather than being silent on supernatural possibilities, negates the possibility of God, or occasions when scientists openly deride those who profess faith in a higher power (such as Richard Dawkins' book *The God Delusion*) bring conflict to the table as well. Respectful communication needs to begin somewhere; a good start would be for individual scientists to stop appearing "patronizing" when it comes to issues of personal faith [38].

A simple but far too common example of a misstep in communication is using religious terms and metaphors to describe scientific experiments. Gary Taubes' unflattering portrayal of particle physicist Carlo Rubbia in *Nobel Dreams* includes a description of the Nobel laureate as "a fundamentalist preacher". This comparison is bolstered by Rubbia's own claim that "The man on the street thinks all matter is made by God in the moment of creation. High energy physicists like us are repeating over and over again the miracle of creation" [39]. Such overreaching finds its way into depictions of fictional scientists as well, painting them in large strokes of hubris and disdain. For example, Maximilian Kohler, the fictional executive director of CERN in the novel *Angels & Demons,* brags "Our scientists produce miracles almost daily" and predicts that science will prove that all deities are "false idols" [40].

In the same novel we also have an example of the opposite problem. Vittoria Vetra explains that her adoptive father, the physicist priest Leonardo Vetra, wanted to "prove Genesis was possible", muddying the waters between religious dogma and the scientific method [41]. Much of the book focuses on the historical tension between science and religion, exacerbated in this work by the introduction of a supposed conspiracy involving scientists in a plot to destroy the Vatican using antimatter. In the real world we also have religious individuals taking it upon themselves to attempt to apply science to religion, akin to trying to fit a square peg into a round hole. For example, self-described biophysicist Frank Lee applies string theory to theology in a series of Internet writings.[3] He claims that the reason why God does not appear in person to those who believe in Him is that He resides in a higher dimension (perhaps taking a cue from Carr and Josephson's work on parapsychology) [42]. At least one well-known mathematical physicist, Frank Tipler, has similarly invoked science to explain theology, much to the vocal disapproval of his colleagues. His 1994 book *The Physics of Immortality: Modern Cosmology, God and the Resurrection of the Dead* was termed "one of the most misleading books ever produced" and a "masterpiece of pseudoscience" by cosmologist George Ellis [43]. Similarly Tipler's 2007 work *The Physics of Christianity* was judged to be "far more dangerous than mere nonsense" by theoretical physicist Lawrence Krauss, because his "reasonable descriptions of various aspects of modern physics, combined with his respectable research pedigree, give the persuasive illusion that he is describing what the laws of physics imply. He is not" [44]. While Tipler certainly has the right to publish whatever he wishes, he is being unfaithful to both physics and religion in producing work aimed at the general public that purports to demonstrate how science "proves" basic tenets of Christianity (or any other faith). It also puts his scientific colleagues in the difficult position of trying their best to demonstrate the error in his application of the science while simultaneously being respectful to the religious adherents to whom Tipler's books are marketed.

It is no wonder, then, that this tension between science and religion is frequent fodder for disaster-based particle physics fiction, with depictions of both scientists and religious individuals often coming off as intolerant and dogmatic in their own ways. Aerospace expert Martin Caidin is certainly less than generous with his depictions of religious individuals in the novel *Star Bright.* Dr. Lawrence Pound, director of the runaway project, is not concerned

---

[3] Most are housed at http://www.geon.us/

about its unexpected self-sustained fusion. He claims it is merely a test from God, whom he believes would never allow the world to be destroyed. He is interrupted by a bullet shot through the window by one of the many equally religious persons who believe that destroying the scientists is actually doing God's will. A religious extremist with the pseudonym El-Asid convinces thousands of followers to attempt to destroy the project, resulting in mass suicides. Their apparent anti-science extremism is complicated by the fact that the electromagnetic effects of Project Star Bright are literally driving people mad, ultimately including Pound. He leaves the facility in order to help El-Asid kill his colleagues, but is instead stoned to death by the mob.

Norman P. Johnson, a former physicist who worked at Argonne National Laboratory and on the Superconducting Super Collider project, describes his religious characters in a particularly negative light in the novel *God's Spark*. Billy Smith, the powerful leader of the Soldiers of God, is a shameless hypocrite. Rather than anti-science by deep conviction, Billy is instead an opportunist, who paints particle physics as the Antichrist in order to serve the needs of the shadowy conspiracy that is behind his rise to international power. He finally begins to question the hold that the conspiracy has over him, and in the end rescues the CERN scientists he had tried to murder. He explains that he has come to realize that the Russians (who have stolen dangerous particles created at CERN) will not be able to control their prize. He is counting on the three scientists he has rescued, experts on the discovery of what he nicknames God's Spark (because of its ability to be a mini Big Bang), to find some means of controlling it and harnessing it for the good of humanity—to do God's work.

The Voices at CERN by Franklin Clermont relies heavily on the stereotype of the Islamic terrorist hell-bent on killing infidels. Dr. Yousaf Khan, a CERN scientist who controls the liquid helium coolant for the LHC's magnets, becomes radicalized due in part to the "God Particle" hoopla, considering it a direct affront to Allah. When the discovery of the afterlife as an 8th dimension is announced to the public, there is speculation in the media that religion is now obsolete. This motivates Khan's plot to destroy the collider during a planned tour of the facility for world leaders in order to punish them for their support of CERN. In the end, CERN tries to heal the rift between science and religion through the formation of an Institute for Non-Theistic Philosophy and Peace Studies, although the term "Non-Theistic" (meaning not involving a belief in a deity) is rather problematic in itself, as it appears to exclude the participation of deists. *The Terror at CERN* gives the reader a view of a facility that is under constant assault by two groups of protestors, one who looks to support organized religion, and the other that just as vehemently seeks to destroy these beliefs. An example is the radical Internet-based EMaCCs, named for $E = mc^2$, who call for the end of all religions. From the 8th dimension the now-dead Khan communicates with his cousin, a Pakistani Colonel, who collaborates with the Christianity-derivative Brothers of the Circle cult to try and destroy CERN with a small nuclear weapon. While this plot is foiled, CERN scientist Roger Fields is nearly assassinated in the restroom while attending his Nobel Prize award ceremony. The overall message of the novel series is that religion is incapable of adapting to new scientific discoveries and there will always be a state of war between science and religion, a less than helpful message, to say the least.

A similar scientific explanation for an afterlife is found in a 1995 episode of *The Outer Limits* "Dark Matters". The interstellar commercial transport vessel *Nestor* finds itself trapped in a closed bubble of space-time, created by a large accumulation of dark matter. Trapped with them are two seemingly lifeless ships, one alien, the other the fellow Earth vessel *Slayton*, thought lost in space with its full crew ten years before. While the crews of the alien vessel and the *Slayton* are dead, they appear as ghosts to the *Nestor's* crew, as their souls are trapped within the closed space-time, existing in "a different quantum state" in this hellish Limbo [45].

A particularly infamous source of tension between science and religion has been the nickname "God Particle" for the Higgs boson, as noted in *The Voices at CERN*. The origin of the unfortunate nomenclature is Leon Lederman's 1993 popular science book *The God Particle: If the Universe is the Answer What is the Question?* He jokingly explains that his publisher wouldn't allow him to refer to the then undiscovered particle as the "Goddamn Particle", but admits in the preface to the 2006 edition of the book that the title succeeded in offending both theists and atheists [46]. The catchy term has taken on a life of its own, showing up in nearly every popular article on the Higgs boson, despite the trouble it causes. The controversy even led CERN to include an explanation for the name in its FAQ [47]. An example of offense being taken with the name can be found in the affidavit of Luis Sancho, who filed a lawsuit against CERN in March 2008 trying to prevent the startup of the LHC. Sancho here complains that CERN is constantly advertising that it "will find the so-called 'God's particle' or Higgs particle, which will explain the origin of mass. This is not truth. God is certainly not a particle" [48]. While the reaction of most scientists reading this is probably a face palm or shaking of the head, Sancho is certainly not the only person to have this opinion. The name is insensitive and has caused unnecessary tension. While it is too late to turn back now, we should learn from the lesson it provides. To those who would argue that I am making too much of this, I will add that the question "Is CERN's aim to prove that God does not exist" had to be added to CERN's social media FAQ [6] (Fig. 4.4).

Fig. 4.4   Leon Lederman (Public domain)

4.5   Beliefs Part 2: Conspiracy Theories

Religion is not the only type of deeply held belief that can complicate an individual's acceptance of particular pieces of scientific knowledge. For some believers, so-called conspiracy "theories" (not true theories in the scientific sense of the word) are articles of faith believed just as fervently as the tenets of any organized religion (although I am certainly not equating the two). Nor does any conspiracy demand (or expect) fidelity, as believers will simultaneously hold several conspiracy theories to be true even when they contradict each other [49].

A real-world example is highlighted in the novel *Angels & Demons*. Reporter Gunther Glick tries to bolster his claim that CERN is run by the Illuminati by pointing out that the CERN logo is a stylized 666, the number associated with the devil in Christian writings. A Google search for "666 CERN" turns up about 693,000 results, including some truly astounding YouTube videos. CERN is very proud of its logo, described as its name and "interlaced rings, which are a simplified representation of the accelerator chain and the particle tracks" [50]. Questions about the logo shape are so common that an explanation of the logo and its adoption is included in the CERN social media FAQ [6]. CERN is certainly not the only entity whose logo has been accused of reflecting some nefarious double meaning. For example, Proctor & Gamble was driven to drop their more than century-old Man in the Moon logo in 1995 due to persistent rumors that the company promoted Satanism (Fig. 4.5).

Such conspiracy theories have dogged CERN for years. A particularly widespread example is the claim that CERN opened up a portal to another dimension in 2016, using an image of a rather vivid storm as evidence. As Snopes.com has explained, the image, taken by Christopher Suarez of a storm over a different part of the Geneva region, was coopted from social media accounts without Suarez's permission [51]. A quick YouTube search for "CERN storm" will turn up videos on the subject with hundreds of thousands and even millions of views.

It is no wonder that conspiracies of various sorts appear in particle accelerator popular culture. In *The Invisible Man* episode "Immaterial Girl" accelerator physicist Lucille McGrier is killed by a colleague because her discovery of how to use the accelerator to achieve fusion threatens the fossil fuel companies that largely fund the accelerator. After the near disaster with a black hole at Filadyne Laboratories in *The Void*, a newscast explains that the Atomic Energy Commission, which had funded Abernathy's dubious project, blames the

Fig. 4.5 CERN logo (User xlibber, CC BY 2.0, via Wikimedia Commons)

accident on a massive electrical fire. Due to the sensitive nature of the research, it is reported that no further information would be released. In *Hyperion*, Dan Simmons' 1989 science fiction tribute to Chaucer's *Canterbury Tales*, one traveler's tale recounts his memories of the release of a small black hole that eventually destroyed the Earth. The apocalyptic event is revealed to have not been an accident, as was widely believed, but rather an intentional interplanetary political act.

A widespread conspiracy involving an unknown number of governments is central to the nineteen-part web series *The Apocalypse Diaries* (2011). The project begins as the video diary of a young couple holed up in their Los Angeles apartment after some unknown catastrophic global event. It expands to become illegal transmissions by a resistance movement fighting against the government conspiracy that seeks to limit public understanding of the incident. Eyewitness footage of the initial incident (suspiciously beginning in Switzerland) features explosions, ear-piercing sounds, and the sky turning orange. Transmissions 15 and 16 are the account of a disguised particle accelerator scientist who obviously describes CERN without naming it. He admits that the initial test of the accelerator at its maximum collision energy caused the event. As in the case of *Star Bright*, the beam became self-sustaining and increases in energy.

Many particle accelerator conspiracies depicted in popular media involve the military applications of accelerator discoveries in some way. In *G.I. Joe: The Rise of Cobra,* scientist Daniel de Cobray is forced against his will to use his civilian particle accelerator laboratory to weaponize warheads containing millions of ravenous microscopic robots that eat virtually anything (perhaps modeled on microscopic black holes). Stephen R. Donaldson's 1994 novel *Chaos and Order* features a space battle between warring factions. Experimental singularity grenades (small artificial black holes launched at ships) considered too dangerous for legal implementation somehow turn up in battle. The supposed discovery of the Higgs boson by Chinese physicists in *A Hole in Texas* leads the Armed Services and Science, Space, and Technology Committees of the U.S. House of Representatives to immediately schedule hearings in order to investigate whether or not the nation is at risk from a boson bomb. Anxiety over the possibility is driven by the existence of a secret project codenamed OMEGA (the last letter of the Greek alphabet being appropriate for discussions of a doomsday weapon).

Sometimes there are so many competing conspiracies in a novel that the reader has trouble keeping them all straight. In David Brin's *Earth* scientist Alex Lustig atones for his creation of several microscopic black holes by working with colleagues to create technology to counteract the real threat, an extraterrestrial microscopic black hole dubbed Beta. Their secret experiments with so-called resonators destroy a dam and send an unsuspecting diver (and accompanying fish) into space, among other collateral damage. There is also the issue of another illegal black hole that ravages the space station, a conspiracy that implicates multiple governments. A renegade set of resonators is used to attack and destroy the scientists' resonators in an effort to control Beta (seeing its potential as a weapon), while there is an even more apocalyptic conspiracy playing out in the novel with the goal of killing all but 10,000 members of humanity.

Pseudoscience beliefs also commonly intersect with conspiracies, especially those widely promulgated on the Internet. An example that seems to have been particularly virulent over recent decades is Nibiru (mentioned in Chapter 3) and the Anunnaki. According to self-proclaimed Sumerian language and culture "expert" Zecharia Sitchin, the ancient Sumerians knew of the existence of a planet-shaped spacecraft named Nibiru piloted by a particularly malicious alien species called the Anunnaki (not coincidentally also the name of a class of ancient Sumerian gods). According to Sitchin, the Anunnaki first genetically engineered and then enslaved humanity, and finally planned to destroy our species through Noah's Flood. Sitchin believed that Nibiru and its ET pilots still pose a threat to our species, as the rogue planet could hit the

Earth and destroy all life on it whenever they wish [52]. Apparently CERN is involved in returning our alien overlords back from whatever parallel dimension they are hiding in, at least if YouTube user gorilla199 aka Chris Constantine (and many a copycat) is correct.[4] Ian Douglas's 1990 *Luna Marine*, Book 2 of his *Heritage Trilogy*, plays with the idea of the Anunnaki as warlike aliens. Ancient Anunnaki technology is discovered on Mars and weaponized on a Moon base occupied by the (here evil) United Nations. This includes a beam of positrons (antielectrons) that is used against American forces. Both the UN and U.S. forces race to develop the Anunnaki's antimatter engine technology first and destroy the other side.

Finally, Patrick Lee's trio of novels *The Breach* (2010), *Ghost Country* (2011), and *Deep Sky* (2012), centers on a wide-reaching conspiracy involving scientists, governments, billionaires, and a secret organization named Tangent. While their stated goal is to understand, guard, and control a wormhole called the Breach and the mysterious advanced technologies that have routinely spit out of it for decades, the truth is far more diabolical. The breach first explosively appears at the initial firing of the fictional Very Large Ion Collider in Wyoming in 1978, and the surviving scientists believe that their experiment is the cause of the catastrophe. However, in the third novel it is revealed that the wormhole is instead an ancient alien structure that is co-opted and intentionally connected to the VLIC by now immortal humans from more than a millennium in the future. Their goal is to influence the 21st century by simultaneously killing 20 million selected individuals with a signal to the brain in order to change history and prevent an Earth-destroying war. While the novel's plot twist clears the creators of the VLIC of culpability (framing them as hapless victims who were largely sacrificed as part of the conspiracy), early press concerning a film treatment under consideration pulls no punches in placing the blame squarely on the Large Hadron Collider (by name). Producer Lorenzo di Bonaventura describes it as "essentially a story about what happens when the supercollider goes wrong" [53].

Conspiracies breed like cockroaches, both in the fictional world of science fiction and our everyday world. While Cook and Lewandowsky warn that we cannot change the minds of fervent believers in such plots, no matter how solid our facts and how flimsy their arguments, this does not mean that we can simply ignore them. Like a cancer, they grow and mutate, spreading across the Internet as they infect countless susceptible individuals. This sets the stage for an analysis of scientists' main communication with the public in recent years concerning particle physics and CERN, their published safety studies.

---

[4] https://www.youtube.com/watch?v=-ugASLblNk4

References

1. F. Clermont, *The Voices at CERN* (CreateSpace, 2014), pp. 8–9
2. F. Clermont, *The Voices at CERN* (CreateSpace, 2014), p. 6
3. F. Clermont, *The Voices at CERN* (CreateSpace, 2014), p. 12
4. National Science Board, *Science & Engineering Indicators 2018* (National Science Foundation, Alexandria, 2018), p. 7.28
5. K. LaCapria, Was a Human Sacrifice Captured at CERN?, Snopes, https://www.snopes.com/fact-check/human-sacrifice-captured-at-cern/
6. CERN Answers Queries from Social Media, CERN, https://home.cern/resources/faqs/cern-answers-queries-social-media
7. J. Cook, S. Lewandowsky, *The Debunking Handbook, version 2* (University of Queensland, St. Lucia, 2012), pp. 1–3
8. J. Cook, S. Lewandowsky, *The Debunking Handbook, version 2* (University of Queensland, St. Lucia, 2012), p. 4
9. National Academies of Sciences, Engineering, and Medicine, *Communicating Science Effectively: A Research Agenda* (National Academies Press, Washington, DC, 2017), p. 64
10. FAQ: LHC the Guide Education, CERN, https://cds.cern.ch/record/2255762/files/CERN-Brochure-2017-002-Eng.pdf
11. A. Kusenko, Viewpoint: Are We on the Brink of the Higgs Abyss? Physics **8** (2015). https://physics.aps.org/articles/v8/108
12. A. Helbing, T. Helbing (script), The Race of His Life, The Flash, season 2 (2016)
13. J. Glassner, D. Levinson (script), Immaterial Girl, The Invisible Man, season 2 (2001)
14. C. Funk, L. Rainie, Public and Scientists' Views on Science and Society, Pew Research Center, http://www.pewinternet.org/2015/01/29/public-and-scientists-views-on-science-and-society/
15. National Science Board, *Science & Engineering Indicators 2018* (National Science Foundation, Alexandria, 2018), p. 7.44
16. M. Goodman, R. Finnegan, L. Mohadjer, T. Krenzke, J. Hogan, *Literacy, Numeracy, and Problem Solving in Technology-Rich Environments Among U.S. Adults: Results from the Program for the International Assessment of Adult Competencies 2012: First Look* (U.S. Department of Education, Washington, DC, 2013), p. 20
17. M. Goodman, R. Finnegan, L. Mohadjer, T. Krenzke, J. Hogan, *Literacy, Numeracy, and Problem Solving in Technology-Rich Environments Among U.S. Adults: Results from the Program for the International Assessment of Adult Competencies 2012: First Look* (U.S. Department of Education, Washington, DC, 2013), p. B-7

18. National Academies of Sciences, Engineering, and Medicine, *Communicating Science Effectively: A Research Agenda* (National Academies Press, Washington, DC, 2017), p. 39
19. J. Carpenter (dir.), Prince of Darkness, Universal Pictures (1987)
20. R.C. Cumbow, *Order in the Universe: The Films of John Carpenter* (Scarecrow Press, Metuchen, 1990), p. 190
21. D. Carey-Hill (dir.), Futurama: Bender's Game, Twentieth Century Fox (2008)
22. B.M. Terhal, M.M. Wolf, A.C. Doherty, Quantum Entanglement: A Modern Perspective. Phys. Today **56**(4), 46–52 (2003)
23. J. Markoff, Sorry, Einstein. Quantum Study Suggests 'Spooky Action' Is Real, The New York Times, https://www.nytimes.com/2015/10/22/science/quantum-theory-experiment-said-to-prove-spooky-interactions.html
24. N. Lambert, Y.-N. Chen, Y.-C. Cheng, C.-M. Li, G.-Y. Chen, F. Nori, Quantum Biology. Nat. Phys. **9**, 10–18 (2013)
25. T. Krantz (dir.), The Big Bang, Anchor Bay (2011)
26. A. Castro, 10 Theories on What Caused the Flash Forward in *Flashforward*, Syfy Wire, https://www.syfy.com/syfywire/theories_on_what_caused_t
27. B. Carr, Is There Space for PSI in Modern Physics?, https://www.tcm.phy.cam.ac.uk/~bdj10/psi/carr2003.html
28. B.D. Josephson, String Theory, Universal Mind, and the Paranormal, ArXiv, https://arxiv.org/abs/physics/0312012
29. M. Durrani, Physicists Probe the Paranormal, Physics World, https://physicsworld.com/a/physicists-probe-the-paranormal/
30. R. Ruyle, The String Theory, a Physical Trace of the Psychical World, in *Proceedings of the 29th Annual Conference of the Academy of Religious and Psychical Research* (Academy of Religion and Psychical Research, Bloomfield, 2004), p. 20
31. G.G. Sparks, C.L. Nelson, R.G. Campbell, The Relationship Between Exposure to Televised Messages About Paranormal Phenomena and Paranormal Beliefs, J. Broadcast. Electron. Media **41**(3), 345–359 (1997); G.G. Sparks, T. Hansen, R. Shah, Do Televised Depictions of Paranormal Events Influence Viewers' Paranormal Beliefs? Skept. Inq. **18**, 386–95 (1994)
32. G. Benford, *Cosm* (Orbit, London, 1998), p. 271
33. G. Benford, *Timescape* (Bantam Books, New York, 1980), p. 235
34. H. Bromley-Davenport (dir.), Xtro II, North American Pictures (1990)
35. J.M.M. Baldwin, *The Collider* (Smashwords, 2012)
36. K. Larsen, This I Believe Understand: The Importance of Banning the B-Word from Science. Astron. Educ. Rev. **6**(2), 118–126 (2008)
37. S.J. Gould, *Rocks of Ages* (Ballantine, New York, 1999), pp. 5–6
38. M. Trodden, Our First Guest Blogger—Lawrence Krauss, Cosmic Variance., http://blogs.discovermagazine.com/cosmicvariance/2005/11/14/our-first-guest-blogger-lawrence-krauss/
39. G. Taubes, *Nobel Dreams* (Tempus Books, Redmond, 1986), pp. xvi–xvii
40. D. Brown, *Angels & Demons* (Pocket Books, New York, 2000), p. 31

41. D. Brown, *Angels & Demons* (Pocket Books, New York, 2000), p. 88

42. F. Lee, Why We Can't See God, http://www.web-books.com/GoodPost/Articles/SeeGod.htm

43. G. Ellis, Piety in the Sky. Nature **371**, 115 (1994)

44. L. Krauss, In Trying to Provide Evidence for Christian Beliefs, a Respectable Physicist Has Bent Science to its Breaking Point. New Sci. **194**, 53 (2007)

45. A. Brennert (script), Dark Matters, The Outer Limits, season 1 (1995)

46. L. Lederman, *The God Particle: If the Universe is the Answer What is the Question?*, rev. edn. (Houghton Mifflin, Boston, 1993), p. 22

47. Frequently Asked Questions, Angels & Demons: The Science Behind the Story, https://angelsanddemons.web.cern.ch/faq

48. Affidavit of Luis Sancho, La Hora Cero, http://lahoracero.org/wpcontent/uploads/2008/09/adnfil20080402_0016.pdf

49. M.J. Wood, K.M. Douglas, R.M. Sutton, Dead and Alive: Belief in Contradictory Conspiracy Theories. Soc. Psychol. Personal. Sci. **3**(6), 772 (2012)

50. CERN Design Guidelines: The Logo, CERN, https://design-guidelines.web.cern.ch/logo-0

51. B. Palma, Clouds Over Geneva Show CERN Opening a Portal to a New Dimension?, Snopes, https://www.snopes.com/fact-check/clouds-over-geneva-show-cern-opening-a-portal-to-a-new-dimension/

52. The Official Website of Zecharia Sitchin, http://www.sitchin.com/; M.S. Heiser, Sitchin is Wrong, https://www.sitchiniswrong.com/nibiru/nibiru.htm

53. A. Chitwood, Lorenzo di Bonaventura Talks David S. Goyer's THE BREACH; Says It Involves an Accident with the Large Hadron Collider, Collider, http://collider.com/the-breach-movie-story-details-lorenzo-di-bonaventura/

5

Searching For Safety in an Unsafe World

5.1 Would I Lie to You? Scientists and Trust

- The anonymous CERN scientist of *The Apocalypse Diaries* finishes his taped confession to the public by noting "what happened was unforeseeable, but also unforgivable. This is a bell we cannot unring" [1].
- In the original *Ghostbusters* (1984), as the titular characters prepare to use their equipment they realize that they have not thoroughly tested it. Dr. Peter Venkman offers "Why worry? Each of us is wearing an unlicensed nuclear accelerator on his back" [2].
- In *Decay* (2012), the low, low budget zombie film directed by then CERN graduate students Luke Thompson and Michael Mazur, fictional graduate students briefly question whether a bioentanglement project at the fictional LHC is safe (spoilers: it's not), but quickly come to the conclusion that CERN wouldn't be doing a dangerous experiment.
- In an interview, Alison Hume, creator of *The Sparticle Mystery*, explains that she got the idea for the series from the "media speculation about what might happen when they switched on the LHC", reasoning that children would like for all the adults to disappear [3].

While speculation that the LHC could banish all adults to a parallel universe or create zombies is clearly unfounded, given the steady diet of mad scientists fed to the public by popular media we should not be surprised that some in the public have wondered if they can trust the record-breaking high-energy experiments at CERN.

© Springer Nature Switzerland AG 2019
K. Larsen, *Particle Panic!*, Science and Fiction,
https://doi.org/10.1007/978-3-030-12206-5_5

Many readers come away from Mary Shelley's *Frankenstein* remembering the spirit, if not the exact words, of Victor Frankenstein's plaintive caution: "Learn from me, if not by my precepts, at least by my example, how dangerous is the acquirement of knowledge and how much happier that man who believes his native town to be the world, than he who aspires to become greater than his nature will allow" [4]. The subtitle of the novel, "The Modern Prometheus", refers to the classic ancient myth of the god who undergoes continual torture after giving fire to humanity. The lessons of Prometheus and Frankenstein resonate in modern society because they reflect a question that cuts across issues of intellectual freedom, public safety, and scientific ethics: are there to be limits set on scientific experimentation, and if so, who sets those limits? Numerous novelists have voiced their opinions on this very question. For example, Michael Crichton's chaos theoretician Ian Malcolm offers in *Jurassic Park* "Science can make a nuclear reactor, but it cannot tell us not to build it. Science can make a pesticide, but cannot tell us not to use it" [5]. In a 1956 draft letter, J.R.R. Tolkien notes that the One Ring is not a metaphor for nuclear energy, but rather of "*Power* (exerted for Dominion). Nuclear physics can be used for that purpose. But they need not be" (emphasis original). He then adds that nuclear physics "need not be used at all. If there is any contemporary reference in my story at all it is to what seems to me the most widespread assumption of our time: that if a thing can be done, it must be done. This seems to me wholly false" [6]. In *Angels & Demons* the Pope's assistant, the Carmelengo, warns that the "language of science comes with no signposts about good and bad. Science textbooks tell us how to create a nuclear reaction, and yet they contain no chapter asking us if it is a good or a bad idea" [7].

Even if we dismiss the trope of the mad scientist as utterly counter to the self-preservation interests of real scientists, what of the well-intentioned foolish or helpless scientist? These gentler villains are also pervasive in popular culture. For example, the uncontrollable magnetic power of the mysterious island in *Lost* is responsible for the so-called "Incident" in 1977, in which scientists building a station meant to harness this energy instead nearly cause a cataclysm. This Swan Station becomes a means to prevent another event by encasing the original site in eight to ten feet of concrete, reminiscent of Chernobyl, the site of the infamous nuclear reactor accident. "How We Lost the Moon, a True Story by Frank W. Allen", a 1999 short story by Paul J. McAuley, features a risky fusion experiment banished to the moon; instead of creating an energy source it forms a black hole that devours our only natural satellite.

While these and myriad other examples are works of fiction, there have been more than enough accidents in the real world to give citizens reason to pause when scientists say something is perfectly safe. Medical science is particularly prone to unintended consequences, simply because the stakes are so high. An especially haunting example is thalidomide, a commonly used treatment for morning sickness in the late 1950s and 1960s in some European countries. American mothers counted their blessings that the drug was not approved for use in this country when it was discovered that it was responsible for causing severe birth defects in thousands of infants [8]. In the case where unavoidable risks exist, scientists do their utmost to minimize them, but human error and unforeseen circumstances still occur. The Centers for Disease Control was rocked in 2014 when it was discovered that at least three cases of human error potentially exposed lab workers (both at the CDC and other facilities) to the Ebola virus, avian influenza, and anthrax [9]. Both the 1986 Chernobyl and 1978 Three Mile Island nuclear power plant accidents were greatly exacerbated by human error [10]. Finally, although the inherent risks of space travel are well understood by astronauts and their families, the tragic losses of the crews of the space shuttles Challenger (1986) and Columbia (2003) were later shown to have been avoidable [11] (Fig. 5.1).

Trust is a delicate thing; once it is broken it is difficult to repair, as the CDC discovered in the wake of the 2014 string of mishaps. In a controversial 2000 paper physicist Francesco Calogero voices concerns about what he sees as the particle physics community's potential "lack of candor" concerning potential risks from high-energy accelerator experiments [12]. In particular, he suggests that some in the particle physics community are "more concerned with the public relations impact of what they, or others, say and write, than in making sure that the facts are presented with complete scientific objectivity" [12]. Calogero explains that nonscientists could become alarmed by reading scientific papers that they cannot understand (easily done since many scientific papers are available online as preprints), or through reading quotations from such papers taken out of context. While unfounded fears could lead to what he calls "'irrational' decision making" that could negatively impact the scientific endeavor, he argues that it is dangerous to try to avoid this through a lack of transparency because it will simply break the public's trust in the end [13].

This potential tension between honesty and self-preservation is reflected in the novels *Flashforward* and *Angels & Demons,* as the fictional directors of CERN in both novels are deeply concerned with mitigating the damage done to their facility's reputation after their respective incidents. The scientists balk at being prevented from admitting their potential culpability, but the directors fear lawsuits that would not only shut down their facility, but also put the

Fig. 5.1 Plaque commemorating the Three Mile Island nuclear accident (User Z22, CC BY-SA 3.0, via Wikimedia Commons)

entire field of particle physics into jeopardy for years to come. Unfortunately, a segment of the real public already questions the trustworthiness of scientists. A 2011 Rasmussen poll of 1000 American adults found that 69% felt that it was at least somewhat likely that some scientists have falsified their data "In order to support their own theories and beliefs about global warming" [14]. The wording of the question could be considered highly leading in its usage of two words that the general public (and too often scientists themselves) use sloppily in relation to science, *theory* (in popular usage meaning a mere hunch or guess) and *belief* (as noted in Chapter 4).

This is the environment of anxiety in which scary sounding high-energy particle accelerator experiments are discussed with the general public. In addition, there is one further complication that must be addressed, the public's general concerns about mathematical risk and uncertainty.

5.2   Risky Business: Probabilities and Uncertainties

As Director Kohler explains in *Angels & Demons*, all progress in science has associated risk. "Space programs, genetic research, medicine—they all make mistakes. Science needs to survive its own blunders, at any cost. For *everyone's* sake" [15]. But what do we do when the planet's very existence is at stake? In *The Krone Experiment* the Russian and American governments agree to jointly built powerful lasers to blast the microscopic black hole that is threatening to destroy our planet. The plan is to release the object's energy and shrink its mass while systematically nudging it out of the Earth and into space. However, there is a known risk that the plan might backfire and bury the black hole permanently within the Earth, where it will gnaw away at the planet's interior. The book's epilogue takes place three years later at the first test of the laser system, and ends with the Russian scientist Korolev saying to the American scientist Runyon "Pray the recoil was in the right direction" [16]. Prayers are certainly no replacement for the mathematics of careful risk assessment. However, who decides which risks are simply not worth taking?

The public wants certainties, whereas science (especially quantum mechanics) deals in probabilities. While science thrives in an atmosphere of competing models and hypotheses, the public potentially sees this as a sign of weakness and indecision. They expect science to be monolithic and decisive, since it is the search for the "truth". This misconception about scientific methodology is not surprising, since our educational system largely rewards students who get the "right" answer on tests and other assignments, especially in math. The public is therefore sometimes confused when science changes its collective mind in light of new and more precise experiments or observations. Sean Carroll explains of his physicist colleagues that both their inherent openness and need to be precise work against them. They admit that there are potentially negative scenarios that are really very unlikely, but not strictly impossible. When communicating this to the general public, instead of saying "'No!' they tended to say, 'Probably not, the chance is really very small,' which doesn't have the same impact" [17]. The authors of the National Science Board's *Communicating Science Effectively: A Research Agenda* caution that because some in the general public know that science carries with it inherent uncertainties, to avoid discussing them is not only dishonest, but runs the risk of having that dishonesty called out, especially if predictions or explanations need to be revised later [18].

There is also the general problem of deciding how much risk (and specifically which risk) is acceptable. Research suggests that people are more comfortable with risks that are familiar and have been part of society for a long time (e.g. smoking cigarettes or drinking alcohol), or that an individual can choose to avoid or take on for themselves (e.g. whether or not to go skydiving). On the other hand, unacceptable risks are those that are not well-known or have effects that only show up years in the future (e.g. side effects of a new drug) or are taken on without consent (e.g. unexpected exposure to chemicals in drinking water) [19].

An often-repeated story of scientists having a rather cavalier attitude towards public risk is referenced by the mad scientist Dr. Thomas Abernathy of *The Void*, when he offers of his own accelerator experiment "with any great endeavor there's always an element of risk. You know everyone thought the hydrogen bomb was going to ignite the atmosphere" [20]. In his popular level book *Collider: The Search for the World's Smallest Particle* Paul Halpern explains that lead scientist Enrico Fermi took wagers as to whether or not the first atomic weapon test (not, as noted above, the later first hydrogen bomb test) would cause the atmosphere to vaporize, what he terms "a macabre kind of gambling". Halpern further offers that the possibility that "an experiment could be conducted with an unknown impact on the fate of the entire planet is shocking" [21].

In an attempt to get to the truth of the matter, John Horgan asked Nobel laureate Hans Bethe to recount his part in the tale during a 2015 interview. Bethe explains that three years before the first atom bomb test, Manhattan Project scientist Edward Teller, later the father of the much more powerful hydrogen bomb, asked if such an explosion could cause the nitrogen in the atmosphere to fuse, mimicking reactions that occur in some stars. Bethe was able to show that this was not a reasonable possibility, and calculations by other scientists showed that other fusion reactions also posed no threat to the atmosphere. When the possibility had originally been brought up, the scientists had properly taken it seriously, including leaders Robert Oppenheimer and Arthur Compton, and had done due diligence in investigating the possible outcomes. Bethe confirms that he and Fermi were "absolutely" certain about the safety of the atmosphere at the Trinity test in 1945, and Fermi jokingly offered the wager only to break some of the tension at the high-stakes test [22]. Yet the urban legend persists, as reflected in questions routinely asked about it on the Internet.

The media also plays a central role in advertising perceived risk, often sensationalizing it in order to boost ratings or readership. Those of us who live in New England suffer through this every winter as television meteorologists turn each potential snowfall into the blizzard of the century. There are more subtle ways that scientists and science popularizers sometimes spice up their stories in order to hold the attention of the general public. An important case study is the March 1999 *Scientific American* article "A Little Big Bang" written by physicist turned science journalist Madhusree Mukerjee. Not only did the title of this article utilize the framing device of comparing the experiments planned for the then upcoming Brookhaven National Laboratory's RHIC (Relativistic Heavy Ion Collider) to the Big Bang, but this analogy is carried throughout the entire article. It is promised that this new collider will "soon create matter as dense and hot as in the early universe" and by using "processes that mimic the big bang—but again are extremely hard to calculate", has the possibility to create not only the quark-gluon plasma, but "innumerable other hypothetical phenomena", including the provocatively named "strangelet: a quark droplet with many strange quarks" [23]. The experiment could even potentially create phenomena "as yet unimagined by theorists" [24].

If the possible results are "hard to calculate" and possibly "unimagined", is it possible that the experiment is patently unsafe? More than one reader thought so and wrote to the periodical to voice their concerns, leading to a discussion in the July 1999 "Letters to the Editor" section. The editor begins by explaining that the article had "alarmed several readers", including one Michael Cogill of British Columbia who questioned the wisdom of attempting to create material that has not existed since the early universe. A letter by Walter Wagner, a lawyer with a B.S. in biology and a minor in physics, poses the possibility that the RHIC could possibly create miniature Hawking black holes. Wagner suggests that such black holes could be "drawn by gravity toward the center of the planet, absorbing matter along the way and devouring the entire planet within minutes" [25]. While Wagner's terrifyingly short timescale for such an event has been shown to be completely off (as will be explained in a later section), such an imminent threat could have terrified readers of his letter to such an extent that his later admission that his own calculations show this would not occur would have been missed. Any relief they might have felt would have been tempered by Wagner's observation that his "calculations might be wrong" [25].

Scientific American decided that the rebuttal should be penned by Princeton physicist (and later Nobel Prize recipient in physics) Frank Wilczek, who had been quoted in the original article. Wilczek acknowledges that all new explorations in science raise questions as to "whether we might unwittingly trigger some catastrophe" (invoking the very same story about Fermi and the atom bomb described earlier), and because of this scientists must take such concerns "very seriously—even if the risks seem remote—because an error might

Fig. 5.2 Frank Wilczek (Kenneth C. Zirkel, CC BY-SA 3.0, via Wikimedia Commons)

have devastating consequences" [26]. He then affirms that RHIC could not create Hawking mini black holes. However, to the chagrin of Brookhaven scientists (and others), Wilczek not only notes that the production of strangelets mentioned in the article is theoretically possible, but raises the possibility that they might grow by "incorporating and transforming the ordinary matter in its surroundings", something he compares to the ice-nine scenario in Kurt Vonnegut's *Cat's Cradle* [26]. Wilczek attempts to calm fears by ending with the comforting thought that strangelets "if they exist at all, are not aggressive, and they will start out very, very small. So here again a doomsday scenario is not plausible" [26] (Fig. 5.2).

Eight years later, CERN chief scientific officer Jos Engelen described Wilczek as "an order of magnitude smarter" than himself, but "perhaps a bit naïve". Engelen admits that the current policy of CERN officials is not to say that the possibility of the LHC destroying the planet is "very small but that the probability is *zero*" (emphasis original) [27]. Mathematician and author of several popular level books Amir Aczel recounts the fallout of the *Scientific American* editorial, including newspaper articles and television news segments hyping the potential danger of RHIC's planned experiments. Perhaps the most poignant and regrettable response was a letter of concern written to Dr. John Marburger, the director of Brookhaven, by a self-described crying eleven-year-old child [28]. No child should be made to fear for his or her life because of sensationalized material in the media that misrepresents science and propagates misconceptions and even pseudoscience. It is for this reason that a number of scientists, myself included, became so deeply involved in debunking the so-called December 21, 2012 Mayan Calendar Apocalypse (as described in Chapter 4). It should be noted that the *Scientific American* article and the resulting fallout did not create the public concerns about Brookhaven's high-energy experiments, which were already described in Gregory Benford's novel *Cosm*, published the year before the article. However, it widely disseminated these concerns and escalated the temperature of the anti-RHIC fever. It also became a teachable moment in science communication—would the Brookhaven staff and other scientists learn the lesson?

5.3 Don't Panic: Safety Studies

In the face of rising public concerns, Director Marburger had a decision to make. It was not wise or feasible to ignore these concerns, so it was decided to address them head on by convening a commission of four scientists from Yale, MIT, and Princeton, including Wilczek, to craft a safety report that would hopefully allay fears. The report was released on September 28, 1999 and addressed three possible doomsday scenarios: the creation of a mini black hole or strangelet, or the transition of our universe into a new true vacuum energy state. The report found that the planned RHIC collisions were not powerful enough to create black holes, the production of strangelets could only occur if they came in doubly unexpected negatively charged and stable configurations, and if it were possible to transition the universe to another state of being, natural processes would have already done so by now. Their conclusion was therefore that there is nothing to worry about [29]. The study was posted online as well as in the form of a peer-reviewed paper the next year.[1]

Similar conclusions were independently reached soon after by three theoretical physicists at CERN who even went so far as to assert that the RHIC experiment would produce no harmful effects in five million years of operation [30]. However, in his criticism of these safety reports, Francesco Calogero points out that there was a potential conflict of interest in CERN scientists supporting the safety of RHIC when the LHC—projected to have far higher energy collisions—was then under construction [13]. Such public debates did not go unnoticed by Hollywood. The Syfy Channel's 2006 original movie *The Black Hole* (directed by Tibor Takács) begins with the following subtitled text:

---

[1] Various safety reports and related scientific papers concerning RHIC and the LHC are discussed in general in this chapter in order to describe the process as it unfolded in the public eye. Specific issues with particular scenarios, such as strangelets, are discussed in Chapters 6 and 7.

> In July of 1999, a panel of nuclear physicists discussed the possibility that a heavy ion collider experiment could result in the formation of a black hole. After an extended debate, the panel decided that such a scenario was not just highly unlikely, but impossible. They were wrong. [31]

The exact science fiction scenario featured in the film will be discussed in Chapter 7; the point to be taken away here is that while scientists wished to put these concerns to bed once and for all, speculative fiction is dedicated to the opposite process, to speculate on the "what if". But it is not just science fiction writers who ask these questions; good scientists do so as well. Black holes had been summarily dismissed as potential problems in the case of the RHIC simply because it would not produce sufficient energy to create them. But this was under the assumption that there only exist three dimensions of space. A 2001 paper by physicists Stephen Giddings and Scott Thomas came to the surprising conclusion that if space has more than three dimensions (for example as predicted by brane models), then black holes could be created at significantly lower energies than previously predicted. In their words, "future hadron colliders such as the Large Hadron Collider will be black hole factories" [32]. After their paper was posted on the public arXiv.org archive in 2000 ahead of peer-reviewed publication, a reporter contacted Giddings, inquiring what would happen if the Hawking radiation mechanism couldn't take care of such black holes. The result was a second paper, "Black Hole Production in TeV-Scale Gravity , and the Future of High Energy Physics", that demonstrated that if such black holes really posed a problem, natural high-energy collisions involving cosmic rays would have already resulted in observable catastrophic events. Giddings also warned the physics community that "journalists regularly read our electronic archives!" [33].

Responding to these new predictions and the potential for renewed public concerns, CERN issued a lengthy safety study in 2003 while the LHC was under construction. All suggested catastrophe scenarios brought up in the RHIC report were revisited in light of advances in theoretical knowledge, and it was acknowledged that if space has more than three dimensions, microscopic black holes might be produced at the LHC. However, the report reaffirmed that the Hawking mechanism would destroy such objects before they could begin to pose a threat. In the techno-speak of the report, "black hole production does not present a conceivable risk at the LHC due to the rapid decay of the black holes through thermal processes" [34].

The accuracy of these safety reports was called into question in a 2004 article by quantum physicist Adrian Kent that pointed out interpretation errors in the risk assessment reports. For example, the CERN scientists' pre-

diction that RHIC would run safely for 5 million years was actually a prediction that the probability of the machine destroying the world in a given year
is 1 in 5 million [35]. Despite the fact that this is a very small risk, the idea
that it was misinterpreted (and misstated) by the authors (who should have
known better) is problematic in terms of engendering public trust. In the
original RHIC safety report the probability of a disaster was ruled to be less
than two ten thousands, which is actually a 1 in 5000 chance, about the odds
of dying of the flu in a normal year and hardly insignificant. In light of the
aforementioned paper by Kent and other criticism, CERN commissioned a
second safety report in 2008 that also came to the conclusion that the LHC
and its experiments did not pose a threat to the planet, largely based, again,
on the fact that cosmic rays carry more energy than a particle accelerator collision, trust in Hawking radiation to take care of microscopic black holes, and
that the harmful variety of strangelets is very unlikely to exist [36].

The result was a flurry of papers confirming and disputing the results, especially those related to mini black holes and Hawking radiation. In a game of
intellectual whack-a-mole, as soon as a new potential danger was brought up
in online discussions, some scientist wrote a paper demonstrating how that
scenario was highly unlikely. For example, it was suggested that the mechanism set up to safely collide the LHC's high-energy beam with graphite (carbon) in the case where the machine needs to be shut down quickly itself posed
a danger because it could cause the same carbon fusion process that fuels some
stars. In response, several CERN scientists did the necessary calculations and
published a memorandum showing that this possibility is "completely
excluded" [37]. Similarly after LHC critic and blogger Alan Gillis [38]
brought up the possibility that the magnets' supercooled helium could create
a so-called bosenova (the explosion of an extreme quantum mechanical
"superatom" called a Bose-Einstein condensate) [39], Malcolm Fairbairn and
Bob McElrath demonstrated that "the Bose-Nova style collapse of ^{4}He is
impossible" [40].

Not only was the opening of the LHC greeted by an avalanche of scientific
assurances of its safety, but an equally vociferous wave of conspiracy theories
accompanied by sensational headlines threatening the imminent destruction
of the planet (such as *The Daily Mail's* "Are we all going to die next
Wednesday?"[2]). There was also the issue of lawsuits filed by opponents of the
LHC, especially Walter Wagner from the *Scientific American* editorial, Spanish
journalist Luis Sancho, and German chemist Otto Rossler [41]. While all of
their lawsuits were eventually dismissed for differing reasons, they continued

---

[2] https://www.dailymail.co.uk/sciencetech/article-1052354/Are-going-die-Wednesday.html.

to widely criticize the project through various Internet platforms, finding kindred spirits in those frequenting conspiracy and pseudoscience YouTube channels and websites. Therefore, while the safety studies done by physicists were intended to allay all public fears, they were not successful in swaying the opinions of a portion of the population, for a number of reasons. For one, as Calogero points out, the specter of conflicts of interest (the fox guarding the henhouse) raises issues of trust. Secondly, the safety reports are written *by* physicists *for* other physicists and are largely unintelligible to the average citizen looking for confidence that the various sensational sounding scenarios are more fiction than science. Thirdly, while science thrives in an atmosphere of debate and revised hypotheses, as previously noted the public seeks certainty while science deals in probabilities. The issues with interpreting and communicating probabilities noted by Adrian Kent certainly didn't add to the public's confidence in the perfunctory claims that the LHC was perfectly safe. There are also things you simply can't control, like human behavior. Whenever human beings are involved, there is the potential for things to go wrong, through bad intentions or simple carelessness. Finally, while scientists do their best to consider every possibility, nature often has other ideas. White dwarfs, muons, the four largest moons of Jupiter, quasars, dark matter, and x-rays are just some discoveries in physics and astronomy that were completely unexpected. Common sense tells you that you can't plan for what you don't expect.

5.4   Expect the Unexpected

CERN scientist Theo Procopides of the novel *Flashforward* explains to his Director that "there was no way we could have known this would happen. There's no expert anywhere who could claim that this was a foreseeable consequence of our experiment" [42]. His colleague, Lloyd Simcoe, echoes this in a press conference, stating of the global blackout "there was no way—absolutely none—to predict anything remotely like what happened as a consequence of what we did. This was utterly unforeseen—and unforeseeable. It was, quite simply, what the insurance industry calls an act of God.... It could not have been prevented". Not surprisingly, the press is not convinced. As one interrupts, "Of course it could have.... If you hadn't done the experiment, it never would have happened" [43].

An unknown event causes an LHC magnet to suddenly lose superconductivity due to a rise in temperature (an incident called a *quench*) in the zombie film *Decay* and the director general appears determined to avoid a more catastrophic event. What his staff does not know is that he is actually ignoring the safety concerns of senior scientists and pushing forward with the all-important search for the Higgs boson. The result is nothing short of a total zombie apocalypse. A magnet incident is also referenced in *The Flash*, as they examine the damaged accelerator ring at S.T.A.R. Labs. Cisco warns that if they don't control the voltage when they try to restart the supercooling system for the magnets "the helium blowback could damage the mountings" [44].

These fictional references to problems with the all-important superconducting magnets makes sense, since they are central to the LHC achieving collision energies higher than other accelerators. But the prevalence of magnet accidents in popular media actually has its genesis in real life. Nine days after the LHC was first turned on in September 2008, a "faulty electrical connection between two of the accelerator's magnets" resulted in damage to dozens of the huge, intricate magnets (tearing several from their mountings) as well as the "release of helium from the magnet cold mass into the tunnel" [45]. CERN Director General Robert Aymar is quoted as explaining that "*This incident was unforeseen*" but CERN can "*ensure that a similar incident cannot happen in the future*" (emphasis original) [45]. While this description from CERN's official press release of the so-called "LHC incident" seems to seriously downplay the accident (which resulted in an unplanned shutdown of over a year), depictions of the event by scientists in their own popularized accounts clearly appear to sensationalize it. Sean Carroll initially states that the LHC "exploded", only to quickly qualify this with "Not the entire accelerator, of course" [46]. Leon Lederman calls it a "*massive cataclysmic explosion*" (emphasis original), explaining how a "human-made bolt of lightning, like a super-bolt from Odin's scepter, blasted through the neighboring magnets and pierced the cryostat.... A mile away steel doors were blown off their hinges" [47] (Fig. 5.3).

Regardless of whatever euphemisms are used to describe this event, the point remains that a small electrical issue mushroomed into a major problem, resulting in significant damage to the facility, but thanks to the safety protocols already in place, no human injuries. Any unforeseen problem has the potential to eat away at public confidence, so while these events cannot be prevented, swift, clear, honest communication is key to operating in the spirit of absolute transparency. Communication that downplays the seriousness of the incident may not achieve the intended consequence of calming fears, especially when scientists themselves describe it in more heated terms in their own popularized versions (done for the sake of reader interest, it is presumed).

A modest measure of honest, self-deprecating, disarming humor can be an effective tactic in deescalating potential fears when responding to truly minor

Fig. 5.3 Superconducting quadrupole magnet for the LHC (Julian Herzog, CC BY-SA 3.0, via Wikimedia Commons)

events. For example, in November 2009 just as the LHC was getting back online after the 2008 magnet incident there was a highly publicized problem with the electrical substation that provided some of the power to the magnets' cryogenic system. When a worker found a piece of bread on critical electrical components at the substation as well as a bread-loving bird, it was widely reported that the smoking gun had been found. CERN Communications Chief James Gillies discounted the cause and effect in an article in the *CERN Bulletin* but playfully referenced the speculation, noting "Headlines about birds and baguettes may be uncomfortable to live with, but it's always worth remembering that this kind of attention is ultimately for the good. Soon, the headlines should be turning from birds to b-quarks, and from baguettes to bosons" [48]. In 2016 a weasel-like animal called a marten[3] came into contact with a CERN transformer and was immediately electrocuted, also causing a

---

[3] The unfortunate creature became part of a museum display (https://www.npr.org/sections/thetwo-way/2017/02/01/512854650/worlds-most-destructive-stone-marten-goes-on-display-in-the-netherlands).

power failure. Scientist Saverio D'Auria, shift leader during the "marten affair", described on the official CERN blog how teamwork got the ATLAS detector back online in short order on what he termed a "difficult day" [49]. All of these events create fodder for science fiction writers, generating a touch of science realism in the eyes of readers and viewers who have heard of one or more of these incidents. For example, a magnet rupture destroys part of an accelerator laboratory in Garfield Reeves-Stevens' novel *Dark Matter*.

There are other potentially unpredictable incidents besides the simple failure of technology, including the discovery of new types of particles or perhaps even new physical principles. Accelerator physicist George Griffin observes an anomalous event he calls a "Snark" (from Lewis Carroll's "The Hunting of the Snark") at *Einstein's Bridge's* fictional SSC, an event that seems to defy the known laws of physics. The puzzling observation is actually an extraterrestrial signal from the species known as the Makers who are trying to contact humanity for benevolent reasons. An editorial on the *Huffington Post* website by author and painter Peter Reynosa suggests that it is exactly these "Unknown Unknowns" that we should be worrying about with accelerator experiments. In his words, the possibility of some dangerous discovery being made at the LHC and other accelerators is an "unthinkable horror that we live with" [50]. While Mr. Reynosa is not a physicist, his concerns deserve due consideration, if for no other reason than they give us a window into an opposing interpretation of what particle physicists view as the exciting possibility of making new discoveries.

Despite every precaution that scientists take, Reynosa argues, you cannot prepare for what you do not know you need to prepare for. A real world example of this involves the unexpectedly large nuclear blasts associated with some of the early hydrogen bomb tests. For instance, on March 1, 1954 the U.S. government's *Castle* Bravo thermonuclear test at Bikini Atoll exploded with a yield of 15 megatons, three times larger than calculations had predicted. The four-mile-wide fireball threatened observation bunkers and fatally irradiated the crew of the Japanese fishing boat the *Fukuryu Maru*—the *Lucky Dragon*. A similar test several weeks later, *Castle* Romeo, yielded three times more energy than expected [51]. Fortunately, as scientific knowledge about nuclear fusion reactions increased we became more proficient in calculating the expected results of our experiments (Fig. 5.4).

In the 21st century the most likely dangerous unknown is not the behavior of subatomic particles, but human behavior, in the form of human-caused accidents and acts of terrorism. This threat shows up quite often in novels and television series. For example, synchrotron accidents and superpowers seem to go hand in hand in popular culture. While the meta-humans of *The Flash* are

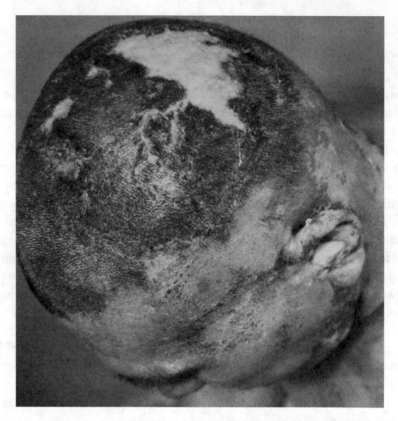

Fig. 5.4 The effects of the *Fukuryu Maru's* accidental exposure to a nuclear test on a crewmember (Public domain)

currently front and center, examples are legion. An accelerator accident involving dark matter bestows superpowers on the unfortunate particle physicist Dr. Chester Banton of *The X-files* episode "Soft Light" (1995). After being accidently exposed to the accelerator beam, his shadow has the power to dematerialize people against his will. Particle physicist Lucille McGrier of *The Invisible Man* episode "Immaterial Girl" is purposefully exposed to the accelerator beam by an evil colleague in order to murder her and completely dispose of the body. In director Sam Raimi's *Spider-Man 3* (2007) escaped prisoner Flint Marko evades the police by trespassing into a particle accelerator facility despite the really obvious "Danger Particle Physics Test Facility" signs. He falls into a giant sand pit just as it is irradiated by some kind of particle beam, transmuting him in the shapeshifter Sandman. Bowerick Wowbagger of Douglas Adams' 1982 novel *Life, the Universe and Everything* is cursed with immortality due to an "unfortunate accident with an irrational particle accelerator, a liquid lunch and a pair of rubber bands" that was never successfully replicated [52].

And then there's Roger, the troublesome alien of the animated series *American Dad!* (2005–). In the episode "The Two Hundred" Roger masquerades as a sixth grader on a school fieldtrip to Langley Quantum Labs' new Hadron Collider on its inaugural day of operation. A scientist points out to the children that they can see inside the collider through a convenient cutout window in the particle ring located only a few feet from the children (rather than buried underground). Roger climbs inside the accelerator tube to retrieve his dropped sunglasses just as the beam starts up. He is pelted with green spiraling lights that represent bosons and his resulting flatulence spreads out like a nuclear blast, causing an apocalypse when his various personas (which had previously appeared throughout the series) "split and form separate physical entities". He later offers to the survivors "I apologize, everyone, I'm the one who ended the world" [53].

The idea of anyone coming into direct contact with an accelerator beam apparently fascinates both writers and the general public. The satirical website *ScienceInfo.news* has a fake news story of a CERN physicist being sucked into a black hole created by the LHC. The story claims that his colleagues continue to create tiny black holes in order to throw supplies to him including "food, water and a flash light" [54]. It is rather telling that several comments to the article question whether or not this is a real story, especially given the website's clear disclaimer. Sean Carroll recounts the story of Anatoli Bugorski, a Soviet scientist who was hit in the face by a 76 GeV proton beam (a mere fraction of the LHC's energy) at the U-70 Synchrotron in Protvino, Russia. His injuries included hearing loss and facial paralysis on his left side as well as scarring due to the radiation[4] [55]. Such an accident is not possible at CERN, due to safety protocols meant to protect staff from the ionizing radiation associated with the underground beam (Fig. 5.5).

While direct human interactions with the beam are ruled out, what about unauthorized activity and theft of material, such as happened with the antimatter in *Angels & Demons*? In the novel, Vittoria Vetra considers the antimatter she and her father create at CERN to be the "ultimate terrorist weapon" and it is used as such by the deeply disturbed Camerlengo [56]. In the *Warehouse 13* episode "Time Will Tell" (2010) a renegade former Secret Service agent uses shapeshifting technology to impersonate a CERN scientist and gain access to an antimatter storage area. A sample of antimatter is secreted out in a metal briefcase and used to power a vest that makes someone move so quickly as to be imperceptible to the human eye. In the episode the security

---

[4] Photographs of his scarring can be found at https://gizmodo.com/what-happens-when-you-stick-your-head-into-a-particle-a-1171981874.

Fig. 5.5 LHC's underground tunnel (Julian Herzog, CC BY-SA 3.0, via Wikimedia Commons)

measure is a retinal scan of two employees with clearance (one up from *Angels & Demons*, where it is a single retinal scan, in that case using an eyeball torn from a murdered scientist). In contrast, the 2009 episode "Pinewood Derby" of the animated series *South Park* paints CERN security as utterly incompetent, when geologist Randy Marsh breaks into the Hadron Particle Super Collider in Switzerland (an obvious play on the LHC) and steals an apparently very small superconducting bending magnet. His Princess Leia costume utterly stymies both the facility security and the local police. More seriously, the grassy berm covering the RHIC in *Cosm* bears bare spots due to gasoline sprayed by vandals, while in David Brin's *Earth* the microscopic black hole Alex Lustig creates in Peru is accidentally set free when an angry mob cuts the power cables to the apparatus keeping the object contained. Alex reflects that he had been "preparing to sabotage the plant myself" before the riot, admitting that scientists didn't understand these "damned things" as well as they thought they did [57].

Given these fictional examples, one can understand why, in a colloquium for the CERN staff, physicist John Ellis notes, "Humanity is safe from the LHC, but is the LHC safe from humanity?" [58]. He elaborates that public replies to Internet stories about potential LHC dangers often include threats against CERN scientists. Likewise, in Franklin Clermont's *The Voices at CERN* threats are made against CERN after the discovery of the 8th dimension afterlife. Clermont writes, "It started with threatening e-mails during the press conference…. Every package delivered since the rumors started received heightened scrutiny in case someone decided to send explosive, biological, or radioactive materials to what was now the focus of world attention" [59]. Threats against particle accelerators is not a new phenomenon, nor are terrorist actions. Two bombs exploded at the Stanford Linear Accelerator Center (SLAC) in 1971, causing slight damage. No suspects were caught, and the exact reason for the bombs was never discovered. One of the hypotheses, that it was an inside job, is echoed in the 2013 episode "Repairs" of *Agents of S.H.I.E.L.D.* Four technicians are killed in a particle accelerator explosion, and while Hannah Hutchins, a quality control engineer, accepts the blame, the accident is actually an act of sabotage by Tobias, one of the dead technicians. He continuously loosened couplings and kept reporting the problem in order to get the attention he craved from Hannah.

Another sabotage scenario is at the center of *12:01* (1993), a made-for-TV movie directed by Jack Sholder. Barry Thomas, an unmotivated office worker at the Utrel Corporation, is the only person aware of the fact that everyone is reliving the same day over and over again in a time loop created by the Utrel particle accelerator. While the accelerator's purpose is to solve the world's energy problem, public outcry at safety concerns has led the government to order the machine shut down. Program director Dr. Thadius Moxley is driven to madness because he had been working on the SSC before it was cancelled and had been narrowly scooped by other scientists on a previous hypothesis. Thomas and Moxley's colleague Dr. Lisa Fredericks discover that Moxley has intentionally sabotaged a seal in the radioactive isotope containment system in order to create a hazmat incident that delays the dismantling of the accelerator. Dr. Robert Denk, the only other high-ranking physicist in the lab, is murdered by Moxley after revealing to Thomas and Fredericks that the Justice Department is investigating Moxley for alleged violations of safety regulations done to keep the accelerator on budget. Fredericks and Thomas interrupt Moxley's desperate late night firing of the accelerator (the event that causes time to lock up in the first place) and when Moxley is hit with his own beam, he and the machine explode, breaking the time loop. The depth of the conspiracy in this film is so extreme that the average viewer should easily brush it off as impossible in the real world. However, can the same be said for those with deep beliefs in conspiracy theories, especially those who already suspect the physicists of RHIC or LHC of putting science before safety?

Dr. Yousaf Khan's lone wolf plot to destroy the LHC in Franklin Clermont's *The Voices at CERN* is to mimic the 2008 accident with the magnets and their coolant, this time involving all of the magnets by creating a computer program to fool the technicians into thinking the system is operating normally until it is too late. His plan is ultimately thwarted after much mayhem, and although he is killed, the CERN administration realizes that Khan can still communicate with the living from the 8th dimension and could inspire someone to try and sabotage the facility again. Terrorism is also central to the plot of the 2009 Syfy Channel original film *Annihilation Earth* (directed by Nick Lyon). The EVE or Electromagnetic Vacuum Energy project generates clean energy by linking three supercolliders at Orleans, Barcelona, and Geneva, the last a direct nod to CERN. Representatives from the oil producing nations of the Middle East (portrayed as blatant and negative stereotypes of Arabs throughout the film) not only object to the fact that they have been left out of this technology, but that their oil is increasingly losing value. When a security breach at the Orleans facility leads to its destruction and the deaths of 20–30 million people, Raja Raheem Bashir, one of the two main scientists on the project, is immediately suspected of terrorism, although it is a known terrorist, Aziz Khaled, who is really to blame. Ironically, it is a bad decision on the part of the European head scientist, David Wyndham, that leads to the destruction of our planet. The film (like *The Voices at CERN*) not only plays on numerous fears surrounding the LHC (and the public's inability to separate fact from fiction), but the rampant fear of terrorism in general, and Middle Eastern terrorism in particular. It is no coincidence that posts to various online LHC protest sites refer to the CERN scientists as "terrorists" bent on destruction.

It is also an interesting coincidence that several months before the premiere of *Annihilation Earth* an LHC scientist was arrested (and later convicted) on suspicion of working with al-Qaida [60]. CERN was quick to clarify that the scientist was not working with any materials that could become instruments for terrorism and that CERN does not work on projects that have potential military applications. It is particularly ironic that this individual was working on the LHCb experiment that seeks to understand the reason why there is so much more matter than antimatter in the universe [61]. It is quite possible that members of the general public reading about this incident could connect it in their own minds with the fictional terrorism using antimatter in *Angels & Demons*, despite the fact that, as we will see in the next chapter, antimatter could not be a plausible weapon. The bottom line is that scientists are human beings, and are susceptible to the same personal foibles as the rest of humanity. We are neither angels nor demons, nor do we possess superpowers or the ability to see the future. We can, however, apply the laws of physics, as we currently understand them, and continue to ethically, honestly, and safely explore the mysteries of the universe. Our next stop will be several specific disaster scenarios suggested by those concerned about particle accelerator research, separating fact from fiction with an eye on how physicists explain their science to colleagues and the general public.

References

1. Transmission Sixteen, The Apocalypse Diaries, http://theapocalypsediaries.com/
2. I. Reitman (dir.), Ghostbusters, Columbia Pictures (1984)
3. Alison Hume *The Sparticle Mystery* Interview, Script Consultant Blog, http://script-consultant.co.uk/2011/03/01/blog-alison-hume-sparticle-mystery-interview/
4. M. Shelley, *Frankenstein or, the Modern Prometheus* (Signet Classics, New York, 2000), p. 38
5. M. Crichton, *Jurassic Park* (Knopf, New York, 1990), p. 314
6. H. Carpenter (ed.), *The Letters of J.R.R. Tolkien* (Houghton Mifflin, Boston, 2000), p. 246
7. D. Brown, *Angels & Demons* (Pocket Books, New York, 2000), p. 477
8. J.H. Kim, A.R. Scialli, Thalidomide: The Tragedy of Birth Defects and the Effective Treatment of Disease. Toxicol. Sci. **122**(1), 1–6 (2011)
9. Laboratory Safety at CDC, Centers for Disease Control, https://www.cdc.gov/about/lab-safety/reports-updates.html
10. A Brief History of Nuclear Accidents Worldwide, Union of Concerned Scientists, https://www.ucsusa.org/nuclear-power/nuclear-power-accidents/history-nuclear-accidents
11. R. Stone, J. Ross-Nazzal, The Accidents: A Nation's Tragedy, NASA's Challenge, NASA, https://www.nasa.gov/centers/johnson/pdf/584719main_Wings-ch2b-pgs32-41.pdf
12. F. Calogero, Might a Laboratory Experiment Destroy Planet Earth? Interdiscipl. Sci. Rev. **25**(3), 198 (2000)
13. F. Calogero, Might a Laboratory Experiment Destroy Planet Earth? Interdiscipl. Sci. Rev. **25**(3), 201 (2000)
14. 69% Say It's Likely Scientists Have Falsified Global Warming Research, Rasmussen Reports, http://www.rasmussenreports.com/public_content/politics/current_events/environment_energy/69_say_it_s_likely_scientists_have_falsified_global_warming_research
15. D. Brown, *Angels & Demons* (Pocket Books, New York, 2000), pp. 121–122
16. J. Craig Wheeler, *The Krone Experiment* (Grafton Books, London, 1986), p. 431
17. S. Carroll, *The Particle at the End of the Universe* (Plume, New York, 2012), p. 191

18. National Academies of Sciences, Engineering, and Medicine, *Communicating Science Effectively: A Research Agenda* (National Academies Press, Washington, DC, 2017), pp. 27–28

19. P. Passell, The American Sense of Peril: A Stifling Cost of Modern Life, The New York Times, https://www.nytimes.com/1989/05/08/us/economic-watch-the-american-sense-of-peril-a-stifling-cost-of-modern-life.html

20. G. Shilton (dir.), The Void, Lions Gate Entertainment (2001)

21. P. Halpern, *Collider: The Search for the World's Smallest Particles* (Wiley, Hoboken, 2009), p. 210

22. J. Horgan, Bethe, Teller, Trinity and the End of Earth, Scientific American, https://blogs.scientificamerican.com/cross-check/bethe-teller-trinity-and-the-end-of-earth/

23. M. Mukerjee, A Little Big Bang. Sci. Am. **280**(3), 63–64 (1999)

24. M. Mukerjee, A Little Big Bang. Sci. Am. **280**(3), 67 (1999)

25. W. Wagner, Letter. Sci. Am. **280**(7), 8 (1999)

26. F. Wilczek, Reply. Sci. Am. **280**(7), 8 (1999)

27. E. Kolbert, Crash Course, The New Yorker, https://www.newyorker.com/magazine/2007/05/14/crash-course

28. A. Aczel, *Present at the Creation* (Crown Publishers, New York, 2010), pp. 208–209

29. W. Busza, R.L. Jaffe, J. Sandweiss, F. Wilczek, Review of Speculative 'Disaster Scenarios' at RHIC, Brookhaven National Laboratory, https://www.bnl.gov/rhic/docs/rhicreport.pdf

30. A. Dar, A. De Rújula, U. Heinz, Will Relativistic Heavy-ion Colliders Destroy Our Planet?, ArXiv, p. 8, https://arxiv.org/abs/hep-ph/9910471

31. T. Takács (dir.), The Black Hole, Millennium Films (2006)

32. S.B. Giddings, S. Thomas, High Energy Colliders as Black Hole Factories: The End of Short Distance Physics, ArXiv, p. 1, https://arxiv.org/abs/hep-ph/0106219

33. S.B. Giddings, Black Hole Production in TeV-Scale Gravity, and the Future of High Energy Physics, ArXiv, p. 2, https://arxiv.org/abs/hep-ph/0110127

34. J.-P. Blaizot, J. Iliopoulos, J. Madsen, G.G. Ross, P. Sonderegger, H.-J. Specht, Study of Potentially Dangerous Events during Heavy-ion Collisions at the LHC: Report of the LHC Safety Study Group, CERN, p. 12, https://cds.cern.ch/record/613175/files/CERN-2003-001.pdf

35. A. Kent, A Critical Look at Risk Assessment for Global Catastrophes. Risk Anal. **24**, 161 (2004)

36. J. Ellis, G. Giudice, M. Mangano, I. Tkachev, U. Wiedemann, Review of the Safety of LHC Collisions, ArXiv, https://arxiv.org/abs/0806.3414

37. R. Assmann, A. Ferrari, B. Goddard, R. Schmidt, N.A. Tahir, Memorandum: Interaction of the CERN Large Hadron Collider (LHC) Beam with the Beam Dump Block, CERN, http://lsag.web.cern.ch/lsag/beamdumpinteraction.pdf

38. A. Gillis, Superfluids, BECs and Bosenovas: The Ultimate Experiment at The LHC, Science 2.0, https://www.science20.com/big_science_gambles/superfluids_becs_and_bosenovas_the_ultimate_experiment

39. Implosion and Explosion of a Bose-Einstein Condensate 'Bosenova', National Institute of Standards and Technology, https://www.nist.gov/news-events/news/2001/03/implosion-and-explosion-bose-einstein-condensate-bosenova

40. M. Fairbairn, B. McElrath, There Is No Explosion Risk Associated with Superfluid Helium in the LHC Cooling System, ArXiv, https://arXiv.org/pdf/0809.4004.pdf

41. E.E. Johnson, The Black Hole Case: The Injunction against the End of the World. Tenn. Law Rev. **76**, 819–908 (2009)

42. R.J. Sawyer, *Flashforward* (TOR, New York, 1999), p. 42

43. R.J. Sawyer, *Flashforward* (TOR, New York, 1999), pp. 127–128

44. A. Schapker, G. Godfree (script), Things You Can't Outrun, The Flash, season 1 (2014)

45. CERN Releases Analysis of LHC Incident, CERN, https://press.cern/press-releases/2008/10/cern-releases-analysis-lhc-incident

46. S. Carroll, *The Particle at the End of the Universe* (Plume, New York, 2012), p. 76

47. L. Lederman, C. Hill, *Beyond the God Particle* (Prometheus Books, Amherst, 2013), pp. 28–29

48. J. Gillies, The Truth about Birds and Baguettes, CERN Bulletin, nos. 47, 48 (2009), p. 2, https://cds.cern.ch/record/1371902/files/2009-47-48-E-web.pdf

49. S. D'Auria, An Insider View of the 'Marten Affair', CERN, https://atlas.cern/updates/atlas-blog/insider-view-marten-affair

50. P. Reynosa, Why Physics Experiments at the Subatomic Level May Cause 'Unknown Unknowns' to Destroy the World, Huffington Post, https://www.huffingtonpost.com/peter-reynosa/why-physics-experiments-a_b_10089660.html

51. R. Rhodes, *Dark Sun: The Making of the Hydrogen Bomb* (Simon and Schuster, New York, 2005), pp. 541–542

52. D. Adams, *The More than Complete Hitchhiker's Guide* (Wing Books, New York, 1989), p. 317

53. B. Cawley, R. Maitia (script), The Two Hundred, American Dad!, season 11 (2016)

54. A Physicist Disappears into a Mini Black Hole Created by the CERN Particle Accelerator, Science Info, http://www.scienceinfo.news/physicist-disappears-mini-black-hole-created-cern-particle-accelerator/

55. S. Carroll, *The Particle at the End of the Universe* (Plume, New York, 2012), p. 87

56. D. Brown, *Angels & Demons* (Pocket Books, New York, 2000), p. 113

57. D. Brin, *Earth* (Bantam Books, New York, 1994), pp. 7–8

58. J. Ellis, CERN Colloquium: The LHC Is Safe, CERN, http://cds.cern.ch/record/1120625?ln=en

59. F. Clermont, *The Voices at CERN* (CreateSpace, 2014), p. 152

60. CERN Physicist Gets 5 Years for Plotting Terror, Science **336**, 656 (2012)

61. N. Clark, D. Overbye, Scientist Suspected of Terrorist Ties, The New York Times, https://www.nytimes.com/2009/10/10/world/europe/10cern.html

6

Pernicious Particles: Subatomic Particles as Villains

6.1 What's the Matter with Antimatter?

In the *Star Trek* universe, the *Enterprise*'s warp engines are powered by the annihilation of deuterium (heavy hydrogen atoms in which the nucleus consists of a proton and neutron) and antideuterium. So-called dilithium crystals (an invented element, as opposed to two lithium atoms bonded together) control the reaction [1]. However, antimatter is also used as a weapon, for example in the *Star Trek: Voyager* episode "Dreadnought". The title refers to a Cardassian self-guided missile made of 100 kg each of matter and antimatter, said to be able to destroy a small moon. Non "Trekkers" may have first come across the term antimatter in Dan Brown's novel *Angels & Demons* or Ron Howard's film adaptation of the same name. In the novel father/daughter scientists the Vetras keep their antimatter experiments at CERN secret, citing safety concerns. Daughter Vittoria explains that while antimatter is an "important technology" that she hopes will ultimately provide limitless, pollution-free energy that could "save the planet", it also has the potential to destroy the planet (in large enough quantities) [2].

The weaponization of antimatter is perhaps its the most common usage in science fiction media, due to the tremendous amount of energy its annihilation with matter creates. For example, in *Angels & Demons*, the largest sample the Vetras create is a quarter of a gram, with the energy equivalence of a 5-kiloton bomb [3]. In Greg Bear's 1987 novel *The Forge of God* malicious extraterrestrials embed tiny super-dense balls of neutrons and antineutrons within the Earth that spiral inward and destroy the planet when they meet. Another extraterrestrial species fights to save a remnant of humanity by

© Springer Nature Switzerland AG 2019
K. Larsen, *Particle Panic!*, Science and Fiction,
https://doi.org/10.1007/978-3-030-12206-5_6

building space arks and terraforming Mars and Venus. In the animated television series *Exosquad* (1993–1994) Phaeton, the paranoid leader of the artificial humanoids called Neosapiens, moves a particle accelerator from France to his lair in Canada in order to make sufficient antimatter to power his doomsday device. If he can't rule the Earth, he reasons, there should be no Earth *to* rule.

It is no wonder, then, that scientists might be especially sensitive to misconceptions and exaggerations concerning antimatter in popular media. Scientists have not only tried to educate the public about the true nature of antimatter (especially the production of antimatter in particle accelerators) but have attempted to allay any fears pop culture misconceptions might cause. For example, in his popular-level book *Antimatter*, physicist Frank Close makes a point of debunking a number of plot points in Brown's novel, starting with claims made in the opening page (which Brown entitled "Fact"). Among the erroneous claims made by Brown (just on this single page) are that antimatter was only recently made at CERN and that CERN would soon be able to create "much larger quantities" of antimatter [4]. Exactly what Brown intends by "much larger" could mean different things to the reader, depending upon whether comparisons are made to the immediately previous sentence, which states that CERN had previously only been able to make a "few atoms at a time" (closer to the truth), or the prior paragraph that screamed out that the annihilation of a gram of antimatter has the equivalent energy of the nuclear weapon dropped on Hiroshima [4]. Close also takes issue with the possibility that antimatter could actually be used as either a commercial source of energy or a weapon of mass destruction, for the same general reason: it's difficult for humans to make antimatter, taking a lot of energy and yielding miniscule amounts of material for all that effort. As Close explains, "there is no possibility to make antimatter bombs, for the same reason you cannot use it to store energy: we can't accumulate enough of it at high enough density" [5]. Despite these realities, Close bemoans the fact that "concerns about antimatter weapons come up in questions at almost every talk that I have given during the last five years" [6].

Close takes advantage of the public interest in antimatter generated by *Angels & Demons*, using it as a teachable moment to instruct his readers about its practical uses. Among these are positron emission tomography imaging, or PET, which uses radioactive sugar molecules that emit positrons (antielectrons) to track activity in the brain and other tissues. Similar techniques can also detect metal fatigue in critical components of aircraft, also potentially saving lives. This love–hate relationship with the science fact and science "fiction" of *Angels & Demons* is evident in myriad debunking articles that focus on the novel and/or film adaptation [7]. Most of these articles parallel Close's complaint that antimatter is suggested as both an efficient commercial energy source and weapon, as well as the fact that Brown has exponentially more antimatter made in the CERN lab than is feasible, either now or in the foreseeable future. These articles feed the public interest in antimatter stirred by Brown's novel while replacing a limited number of sensational misconceptions with an equally small number of fascinating truths (such as actual uses of antimatter). In this way, some of the best practices noted in Chapter 4 are embraced: emphasizing key facts while avoiding the backfire effects such as overkill; flagging any mention of the misconceptions/falsehoods with a clear, explicit warning; and clearly and succinctly explaining why the misconception is wrong.

When director Ron Howard decided to film part of his adaptation of Brown's novel at the LHC (despite the fact that the LHC is not the facility at CERN that produces antimatter), CERN in turn took the opportunity to extend their partnership with the studio in developing a web presence based on the film called *Angels & Demons: The Science Behind the Story.*[1] We will take this opportunity to analyze the website's use of best practices (or not) in telling the story of antimatter, debunking misconceptions, and heading off potential fears. The main page summarizes the plot of the film and asks several framing questions: "But what is antimatter? Is it real? Is it dangerous? What is CERN?" From here links take you to an overview of CERN (including its mission, structure, and a summary of the LHC), antimatter (including how to make and trap antimatter), frequently asked questions (FAQ), and videos.

The videos focus on the Higgs particle, the LHC in general, and various aspects of antimatter. Some of the titles themselves directly address common misconceptions and potential fears, such as "Does CERN produce antimatter?" and "Could antimatter be stolen?" These two brief (roughly a minute each) videos feature CERN antimatter physicist Rolf Landua, a credentialed expert, succinctly and clearly answering these questions. In the first, Landua literally invites the viewer to look at the huge magnetic apparatus where antimatter is trapped, clearly too large to steal (even by the Illuminati). He firmly explains that small antimatter traps such as those in the film might be possible in the future, but would hold much less antimatter than suggested in the film. In the second video Landua specifically notes that in the entire history of CERN only a few ten billionths of a gram of antimatter had been made. A longer 3-minute video "Why study antimatter" features antimatter researcher Michael Doser, another credentialed expert, directly debunking the danger of

---

[1] https://angelsanddemons.web.cern.ch/

antimatter. He appears to falter a bit when he awkwardly notes that "antimatter, in principle, is absolutely innocuous, if you don't have a lot of it. If you have much more of it than we could ever possibly make then it might start getting a little bit dangerous." But he hits the right notes when he then confidently compares the danger (and energy equivalency) of the amount of antimatter we can currently make with that of a 5 W lightbulb running for a few seconds, an analogy the public should have little trouble relating to [8] (Fig. 6.1).

Most of the individual pages concerning CERN and antimatter are short, clearly written, and make confident statements meant to both calm fears and pique interest. For example, the "Making Antimatter" page features an x-ray of a human skeleton with the statement "*Your body emits antimatter!*" (emphasis original), due to the natural decay of the potassium-40 taken up by the body through food, water, and respiration [9]. Unfortunately, other pages lack explanatory diagrams or other visuals (as suggested by Cook and

Fig. 6.1 Antiproton Accumulator and Antiproton Collector at CERN (CERN, CC BY 4.0, via Wikimedia Commons)

Lewandowsky). For example, the page on trapping antimatter has a photograph of a Penning trap but does not have a diagram explaining how it works. The most puzzling graphics choice is on the main antimatter page, which features Dirac's 1928 equation that predicted the existence of antimatter. Not only is the equation completely alien to the nonscientist, it is presented with no introduction or explanation, and therefore adds little to the educational value of the page.

The FAQ section hits all the necessary questions, but presents them in a rather curious order. "Does antimatter exist?" would logically be the first question addressed, ahead of "Can antimatter be used as an energy source?" The next two topics "Does CERN conduct secret research?" and "Does CERN create black holes?" seem to interrupt the flow of the page and delay the answering of two more relevant questions, "How is antimatter contained?" and "Is it possible to make a bomb out of antimatter?" The page finally ends with "What does the name 'CERN' mean?" and "What is the God particle?", questions that, while not directly related to antimatter or the film, are common enough queries concerning CERN to deserve inclusion here. The individual FAQ pages are generally concise, clearly written, and focused on debunking one specific misconception without falling victim to backfire effects. On the "Can antimatter be used as an energy source?" page the statement that all of the antimatter ever created by CERN could only power a single light bulb for "a few minutes" is illustrated with a photograph of a simple light bulb and the caption "Antimatter is not a viable source of energy" [10]. The juxtaposition of the visual with the confident statement of fact are an effective presentation of this important information. The answer to the question "Does CERN conduct secret research" is immediate and emphatic: "CERN does not conduct any secret research" [11]. Those reading further receive more information about what CERN does do, but this simple statement addresses the main point.

The answer to "Is it possible to make a bomb out of antimatter?" is more drawn out, as this page first notes that antimatter/matter annihilation "could theoretically be used in a destructive way". This distracts the reader from the main point of the second paragraph, that "there is no way" to accumulate enough antimatter to make a bomb. The more detailed explanation of "How is antimatter contained?" may confuse some readers, but the screen capture from the film with the caption "Portable antimatter traps, as seen in the movie, are not feasible in reality" should stem most anxiety (although it is wondered if "feasible" is the best word choice here, rather than "achievable" or "reasonable") [12]. Despite these minor criticisms, the *Angels & Demons* webpages at CERN demonstrate how science can collaborate with the media to educate and entertain the public in a meaningful and respectful way.

6.2   Ghosts in the Machine: Neutrinos

Antimatter doesn't corner the market for villainous particles in science fiction. In the 2013 pilot episode of the adult animated science fiction sitcom *Rick and Morty* misanthropic (and intoxicated) mad scientist Rick Sanchez takes his grandson Morty Smith on a late-night ride in his space cruiser. Rick has decided to give our planet a fresh start by wiping out the entire human species with a neutrino bomb, with the exception of himself, Morty (the new Adam) and Morty's friend Jessica (Morty's Eve). While the bomb does not go off, in the third season it is revealed that Morty has learned how to defuse his grandfather's bombs, which are apparently not only unreliable but more numerous than suggested in the pilot episode.

Neutrinos did not appear as a potential threat in the LHC safety studies for one very good reason—they pose no risk because they rarely interact with matter. Nonetheless, they are often painted as spooky because of their behavior, including their ghostly ability to literally pass through walls of solid lead trillions of miles thick [13]. In their classic 1934 paper "The 'Neutrino'" Hans Bethe and Rudolph Peierls note that a neutrino could pass through the Earth "like a bullet through a bank of fog" [14], while John Updike's poem "Cosmic Gall" compares their effortless travels through our planet to that of "dust-maids down a drafty hall" [15]. There is another reason for the neutrino's dubious reputation; they are produced in large numbers in the supernova explosions of massive stars, serving as celestial harbingers of doom that reach our massive neutrino detectors hours before the actual explosion of the star is visible in our telescopes. In fact, supernovae can be thought of as "neutrino bombs" [16]. Perhaps this is why, in the *Star Trek* universe, Bajoran wormholes are said to give off elevated numbers of neutrinos whenever something passes through them [17] (Fig. 6.2).

Therefore, if you want to include a particle in your science fiction novel or film that does counterintuitive things, a neutrino seems to works just fine. In Greg Bear's 1998 novel *Foundation and Chaos* (based on Isaac Asimov's *Foundation* series), the Three Laws of Robotics, which prevent robots from allowing humans—or themselves—to be harmed, are erased from a robot's positronic brain after being exposed to a neutrino storm. Neutrinos also trigger the end of the world in the apocalypse blockbuster *2012*. Here an abnormally large storm of mutated neutrinos is unleashed from solar flares.

Fig. 6.2   Clyde Cowan and Frederick Reines proved the existence of the neutrino at the Savannah River Plant in 1956 (Public Domain)

In violation of the known laws of physics, these mutant neutrinos heat up the Earth's core and create the impossibly large tectonic shifts that are featured throughout the film. Finally, in the novel *Flashforward* the cause of the global blackout is tied to neutrinos from Supernova 1987a interacting with the LHC. Neutrinos might not interact that often with matter, but science fiction seems to think that when they do, it's very bad. If the connection with supernovae isn't the stuff of nightmares by itself, neutrinos were recently discovered to be connected to another generally unpleasant cosmic scenario, the overactive supermassive black holes at the centers of active galaxies. Recall that the sources of most cosmic rays are still not definitely known. If a recent study based on the observation of neutrinos with energies over 30 TeV holds up, we might have a smoking gun connecting these ghostly particles with the fearsome engines of quasars and other hyperactive galaxies as well as cosmic rays [18]. I suspect that some science fiction writer somewhere is currently writing this into a screenplay, regardless of how the science pans out.

6.3   Fear of the Dark (Matter)

Philip Pullman's *His Dark Materials* trilogy of children's fantasy novels centers around the mysterious real-world material called *dark matter*. Termed "dust" in Pullman's fantasy, he imbues it with a consciousness and involves it in a complicated relationship with human intelligence, parallel universes, and theology. Pullman explains that he found the very name dark matter "evocative, mysterious and profound" [19]. It is no wonder, then, that dark matter appears in different forms (most of which tinged with at least some sinister properties) in many science fiction media.

Dark matter is both an important fuel source and an environmental hazard in the world of the adult animated sci fi comedy *Futurama*, serving as a caricature of fossil fuels (most notably oil). The episode "The Birdbot of Ice-Catraz" is a satire of the Valdez oil tanker disaster. The space supertanker Juan Valdez, filled with "rich Colombian dark matter" spills its cargo across a penguin preserve on Pluto, covering the birds in what looks like black liquid [20]. The dark matter has the unexpected effect of making the penguins super-fertile, with even males laying hastily hatching eggs, leading to an environmental disaster. In the franchise's direct-to-video film *Bender's Game* the dark-side (pun intended) of the commercial dark matter fuel monopoly is exposed. It is discovered that Nibblonians, diminutive intelligent extraterrestrials that defecate dark matter nuggets, are caged and force-fed in mines in order produce the valuable fuel source. Professor Farnsworth had caused the problem in the first place, three decades before, when he created a meta-particle that had simultaneously changed all dark matter into its fuel form (similar to Vonnegut's ice-nine or a strangelet). He corrects his error by bringing the meta-particle into contact with its anti-particle, changing the dark matter back into its useless inert form.

Dark matter is mutated into a harmful form in the *Star Trek: Voyager* novel *Cloak and Dagger* (the first of the three volume *Dark Matters* series [2000] by Christie Golden). An evil member of the primordial Shepherd species uses the Romulan Telek's wormhole technology to generate the harmful dark matter, which infects both the machinery of the starship *Voyager* and its crewmembers (giving them both physical and psychological ailments). At the end of the third novel *Shadow of Heaven* (2000) Telek leaves behind a written statement explaining that dark matter "exists simultaneously in *all* universes" and the Shepherds had taken it upon themselves to balance out the amount of dark matter manifest in each universe so that they will all evolve in the same way [21]. In the Japanese animated film *Harlock: Space Pirate* (2013), directed by Shinji Aramaki, dark matter fuels the engines of the titular character's renegade ship, a technology only possessed by the nearly extinct Niflung species. The very same uncontrollable dark matter had been previously unleashed by Harlock in an effort to make a protective shield around the Earth, but instead it had rendered the surface of the planet unlivable for years and made Harlock immortal. Finally, a dark matter-producing particle accelerator experiment creates the mutant meta-humans in *The Flash*. When a similar incident is intentionally recreated to give Barry back his Flash superspeed, stray dark matter reanimates a dead meta-human in the morgue, creating a kind of meta-zombie.

Dark matter itself is rather zombie-like in some ways. It is not normal matter (not composed of protons, neutrons, and electrons) but still affects the behavior of normal matter. The history of dark matter research dates back to the 1930s, when astronomer Fritz Zwicky studied the Coma cluster, a rich concentration of thousands of galaxies. He found that if you estimate the amount of matter in each individual galaxy and add it all up there is no reason for the cluster to exist—it should have been torn apart by the expansion of the universe long before now. He suggested that there had to be some invisible material whose gravity is helping to hold the cluster together. Studies of the motions of stars around the centers of spiral galaxies like the Milky Way (especially research done by Vera Rubin and collaborators) have also suggested that most of the matter in an individual galaxy is made of this mysterious dark matter as well. Despite the fact that this utterly invisible material outweighs normal matter by a factor of about 5–1, identifying its exact nature has eluded physicists for decades. Evidence has ruled out (or rendered highly unlikely) some candidates but has confirmed none of the rest. One of the hopes for the LHC is that it will find candidate particles or further limit the field of contenders (Fig. 6.3).

The odds-on favorite candidates for dark matter are the WIMPs, weakly interacting massive particles. With predicted masses between 1 and 1000 times that of a proton, the collective gravitational pull of these particles would

Fig. 6.3 Coma cluster of galaxies (ESA/Hubble, CC BY 4.0)

have a profound effect on structure in the universe (such as galaxy clusters). But until these hypothetical particles are actually discovered in the laboratory they cannot claim their crown. Another possibility is the axion, another hypothetical particle proposed to solve a particular symmetry-breaking problem with the strong force. Other speculations involve Kaluza–Klein particles that arise in models with additional dimensions of space. This extradimensional dark matter would largely reside in the higher dimension, which would explain why we have had such a hard time detecting it (except through its gravity) [22].

One particular WIMP candidate has been frequently mentioned as the most likely answer to the problem. If supersymmetry does exist (see Chapter 1), one of the many proposed supersymmetric particles might be the dark matter. Photinos (the SUSY partner of the photon) play a sinister role in *Ring* by Stephen Baxter. Lieserl is an artificial life form designed to survive inside the Sun in order to find out why our star is evolving thousands of times faster than it should. She finds a flock of lens-shaped, sentient photino life forms that she dubs photino birds. She reasons that these creatures feed off the steady heat of stars, and are intentionally converting all current stars to safe, predictable white dwarf stellar corpses in order to avoid dangerous supernovae that would disrupt their nests.

While *Ring*'s hard science fiction is specific in its identification of dark matter, its exact nature is left vague in much of popular media. In the mythology of director Alan Taylor's *Thor: The Dark World* (2013) the leader of the Dark Elves had created a substance called Aether, a weapon of mass destruction, out of the primeval darkness. Astrophysicist Dr. Jane Foster, Thor's love interest, comes into contact with Aether and learns firsthand that it converts matter into dark matter (depicted as a rather negative experience). Dark matter also has a villainous effect on the human body in director Jason Bourque's 2006 made-for-television film *Dark Storm*. Six years after a catastrophic accident involving a dark matter extraction and containment experiment destroys his lab in Montana, mad scientist Dr. Sever McKray wisely places the dangerous dark matter containment process on board an unmanned satellite. A floating bolt unexpectedly pierces the satellite's containment module, releasing dark matter on the same day that General Killion is at the lab to witness a demonstration of the dark matter's potential to disintegrate matter. Lone voice of reason and noble scientist Dr. Daniel Gray urges postponing the test due to safety concerns, but McKray overrides him, and Daniel is contaminated with dark matter. Killion shuts down the project when freak electrical storms cause power outages and disintegrate a spy plane, but not before a huge cloud of dark matter grows in the atmosphere, just as Dan's models had warned.

As one might expect, Dan develops strange superpowers after his dark matter exposure, for example, having the ability to contain small balls of dark matter inside an electromagnetic field and wield them like deadly baseballs. McKray steals the dark matter core from the facility (having sold the weaponized technology to a foreign interest) and kidnaps Dan, while the dark matter storms wreak havoc across the globe. McKray reveals that he has purposefully exposed Dan to dark matter in order to study the effects on human tissue before intentionally exposing himself as well. After descending into cartoonish megalomania, McKray is overpowered by Dan and his friends, and the dark matter is safely sucked out of McKray before he is disintegrated by the machine.

Not only do dark matter's still unknown identity and properties make it a convenient foil in science fiction, but as noted by Philip Pullman, its very name adds to its mystery. In her popular-level work *Dark Matter and the Dinosaurs,* physicist Lisa Randall describes conversations with both a game designer and a screenwriter interested in dark matter's potential as an energy source (presumably in order to exploit it in their respective fictional media). A marketing professional directly questions Randall about the wisdom behind the name, given the bad juju associated with anything dark (e.g. black cats, dark spirits, and dark storm clouds). Randall herself offers that the name leads to misconceptions that dark matter and black holes are the same thing [23]. The lesson of dark matter is that, as in the case of the Higgs "God" particle,

Fig. 6.4 An artist's rendition of our galaxy's dark matter halo (ESO/L. Calçada, CC BY 4.0, via Wikimedia Commons)

words matter; the more mysterious something sounds, the wider the door opens for misleading media representations, something scientists would be wise to take into account when glibly coining terminology in the future (Fig. 6.4).

6.4   Killer Strangelets

Calling something dark matter definitely burdens it with a negative stereotype. The same is true (if not more true) with strangelets. Remember that physicist Frank Wilczek had compared strangelets (dense balls of quarks, including some strange quarks) with the fictional ice-nine of Kurt Vonnegut's novel *Cat's Cradle*. Dr. Felix Hoenikker's fatal final project in the novel had been in response to repeated pestering by a Marine general to create "a little pill or a little machine" that would instantly solidify mud and make it easy for Marine maneuvers [24]. Felix names his solution ice-nine, and his colleagues think it to be hypothetical (like strangelets). Unfortunately, it is far too real and causes the end of the world by converting all of our planet's water into ice-nine.

Recall that in the aftermath of the *Scientific American* editorial that had initially raised the specter of strangelets, Wilczek and three colleagues had written a safety report for RHIC, including an analysis of the potential dangers from strangelets [25]. Not only was it noted that there is no evidence for the existence of strangelets, but calculations suggest that small lumps of strangelets are unstable. Finally, the only type of strangelet that could potentially get larger would be one with a negative charge (because it would attract positively charged nuclei made of protons and neutrons). However, negatively charged strangelets require the addition of more strange quarks (because they contribute a negative charge) and these are more difficult to make because they are heavier than the normal up and down quarks that make up protons and neutrons. Wilczek and his colleagues determined that the chance of producing a dangerous stable, negatively charged strangelet in the entire lifetime of the RHIC was about 1 in 10^{21} [25], conclusions that were similar in spirit (if not exact probability) to those found by other safety studies [26]. Finally, it was found that it is actually more likely that potentially dangerous strangelets would be created at lower rather than higher energies. Therefore since RHIC never made them, LHC wouldn't make them either.

The short-lived television series *Odyssey 5* (2002) directly dealt with dangerous strangelets. In the episode "Trouble with Harry", an evil and insane artificial life form named Phaedra attempts to destroy our planet with strangelets and must be stopped by the crew of the titular space shuttle. In a detailed exposition scene, strangelets are characterized as "piggy" as they "never know when to stop eating" [27]. While it is noted that they hadn't been found at Brookhaven or CERN (mentioned by name in the episode), a fictional Texas supercollider (modeled on the SSC) was planning to intentionally produce strange quarks. Collider physicist Dr. Janice Kitaro is 100% sure that her experiments with the quark–gluon plasma are perfectly safe and will only make "essentially harmless" positively charged strangelets [27]. Her hubris is deflated when the crew demonstrates (in the nick of time) that Phaedra has added code to the computer system that will change the experiment to create dangerous negative strangelets.

Again, it isn't only mad scientists who are seen to pose a threat to the world in these works, but helpless, foolish, and foolishly confident scientists as well. A strangelet-like material recurs in the Marvel superhero universe shared by the television series *The Agents of S.H.I.E.L.D.* (2013–) and *Agent Carter* (2015–2016), among others. Zero Matter or Darkforce is a dangerous unstable material from another dimension that sucks energy out of anything it encounters. For example, in the 2016 *Agent Carter* episode "The Lady in the Lake", a woman's body is found in a frozen block of strange material in an icy Los Angeles lake in the middle of a summer heatwave (reminiscent of ice-nine). Strategic Scientific Reserve Agent Peggy Carter reasons that the victim worked at a particle accelerator because her body glows due to exposure to uranium (pandering to a common misconception regarding radioactivity). The victim is Isodyne Energy Lab particle physicist Jane Scott, who had come into contact with Zero Matter. A final comparison can be drawn with some of Dan's dark matter-created superpowers in *Dark Storm*. These include the ability to freeze water on contact, just as dark matter streaming from the satellite towards Earth causes freezing overnight temperatures. In the film Dan predicts that the growing dark matter storm is attracting all dark matter to itself and incorporating it into the cloud, just like ice-nine or a dangerous strangelet.

Unfortunately for fans of sinister strangelets, a 2018 paper suggests that it is much harder to make highly strange (negative) strangelets than originally thought, so sleep well tonight—no negative strangelet piggies are going to gobble you up any time soon [28].

6.5 Boson Bombs and False Vacuums

Given the importance of the Higgs boson to the Standard Model and all the very public hoopla over its discovery at LHC, it is inevitable that it would be adopted by writers as yet another threat to the planet. Long before its discovery at the LHC, the Higgs boson played a guest role in the final season of the quirky sci-fi series *Lexx*. The crew of the *Lexx* arrives at Earth and find its residents to be rather unintelligent and belligerent. While the captain, Stanley Tweedle, wants to blow up Earth, the robotic head 790 notes that type 13 planets like ours usually destroy themselves at this stage of technology, through war, environmental crisis, or most commonly by being "unintentionally collapsed into a pea sized object" through a search for the Higgs boson [29]. In the 2002 series finale "Yo Way Yo", the self-serving scientist Dr. Ernst Longbone builds a spaceship ark for himself and a hand-picked bevy of beautiful young female technicians when Earth is threatened by an asteroid collision. He gives his other colleagues the consolation prize of a new ultra compact particle collider that will allow them to discover the mass of the Higgs boson before the experiment collapses the planet (as previously predicted by 790).

The mad scientist CERN director in the graduate student made zombie film *Decay* arranges for the death of Dr. Niven, a colleague who argues for a delay in the search for the Higgs boson over safety concerns. Dr. Niven and all his team are intentionally burned by so-called Higgs radiation but reanimate as zombies. Perhaps the best/worst line of dialogue in the film occurs when it is discovered that the Higgs radiation kills the brain except for the brain stem: "The Higgs killed them all" [30]. It is not zombies but evil aliens that threaten particle physicists (and the world at large) in *Einstein's Bridge*. In the novel the search for the Higgs at the SSC leads to an invasion by hostile extraterrestrials from another universe. To save the world, two particle physicists travel back in time and prevent the collider from being built in the first place. As clearly fictional as this might sound, physicists Holger Bech Nielsen and Masao Ninomiya suggest that the reason why the SSC was cancelled (and the LHC had initial troubles such as the magnet incident) is because the universe conspires to prevent the Higgs particle from being discovered [31]. If so, the joke is on the universe, because the Higgs was discovered in 2012 at the LHC and we're still here (Fig. 6.5).[2]

Into the Looking Glass by John Ringo, a convoluted and often parodic tale, straddles the boundary between hard science fiction and scientific terminology salad (such as "We may be able to adjust one of the inactive bosons to form a stream of unique quarks") [32]. Former MIT scientist Ray Chen takes a position at the University of Central Florida because he prefers Florida weather and tries to discover the Higgs particle using a small accelerator with a 4 meter wide ring. In addition to the Higgs boson, he generates a 60 kiloton

---

[2] Again, if, you want to confirm for yourself, see http://hasthelargehadroncolliderdestroyedtheworldyet.com/

Fig. 6.5 Graduate students discover the effects of "Higgs radiation" on the brain in *Decay* (2012) (H2ZZ Productions, CC-BY-NC; screen capture by author) (Public domain)

explosion and an invasion of Earth by hostile aliens from another universe who use Higgs particles as cross-dimensional gateways. Extraterrestrial intelligences are connected with the Higgs in a different way in the 2002 remake of *Solaris*, directed by Steven Soderbergh. Astronauts are sent to a space station orbiting the turbulent alien planet Solaris to determine if it is commercially viable as either property or an energy source. Most of the crew die, driven mad by the appearance of doppelgangers of their loved ones on the station ("visitors" created by the planet). Scientists on the station build a miniature proton accelerator and "matter-phase modulator" to create a beam of anti-Higgs bosons by generating a "negative Higgs field" in order to disintegrate the visitors [33]. The tradeoff is that using the device causes Solaris to grow in mass, resulting in the space station being pulled into it.

Concerns over the search for the Higgs at a fictional Texas supercollider in *The God Particle* are not due to any potential danger from the Higgs boson itself, but rather Steve Keeley's increased ability to manipulate matter when he is near the machine. In particular, physicist Mike McNair worries that a slight change to one of the physical constants of nature by Steve could lead to the destruction of the entire universe. While this is clearly science fiction, it does mirror concerns voiced in some of the LHC safety studies that were resurrected thanks to comments by Stephen Hawking. In his Foreword to the book *Starmus: 50 Years of Man in Space*, a collection of lectures by scientists and astronomers, Hawking notes that the Higgs potential has the "worrysome [sic] feature" that it could undergo a sudden phase transition to a true vacuum at energies above billions of TeV [34]. This possibility, visited in brief in

Chapter 1, would result in catastrophic changes in our atoms and the destruction of the universe as we know it. While this energy range is far beyond the 13 TeV of the LHC, this fact could be easily overlooked by the casual reader, especially with Hawking proclaiming that such a deadly event could completely catch us unawares at any moment (although, he admits, "the estimated lifetime to decay is larger than the age of the universe") [34]. Hawking's comments were widely reported in the media, and raised concerns in the physics community that the public might take his sensational comments the wrong way. Again, various safety studies for both RHIC and CERN had concluded that the chance of this happening is so tiny as to be insignificant (1 in about 10^{36} for the case of RHIC) [35]. John Ellis, one of the authors of the 2008 CERN safety study, reframed the conversation in the media from one of risk to exciting evidence that there is still "new physics yet to be discovered" [36]. In the CERN Colloquium "The LHC is Safe", where he debunks safety concerns for an audience of his colleagues, Ellis clarifies that the LHC will not cause a transition to a true vacuum, but could instead give us information about whether or not the true vacuum exists [37]. Again, there is bad news for doomsday devotees, as a 2018 paper suggests that the lifetime of our current vacuum is at least 10^{58} years and perhaps as long as 10^{139} years, far longer than the lifetimes of all the stars that will ever exist in our universe [38].

If the Higgs won't destroy the universe, could it at the very least destroy the world? This scenario is the basis for one of the most over the top particle physics disaster films, *Annihilation Earth,* introduced in Chapter 5. Set in 2020, this Syfy Channel original film begins with a ghostly cloud travelling along the underground tunnel of a huge particle accelerator modeled on CERN. The cloud impacts a target in a flash of light, and the scene morphs to scientists in radiation suits carefully walking through the ruins of a large city, the inferred connection crystal clear to the audience. Events in the movie are revealed to take place along a timeline beginning some 80 hours before "extinction", according to the subtitles that occasionally appear on the screen. United Nations representative Paxton informs head scientist David Wyndham that security at the Orleans, France accelerator (one third of the EVE or Electromagnetic Vacuum Energy project) has been breached. Wyndham's friend and scientific colleague Raja Raheem Bashir has moral quandaries about their secret development of a simulation called the Doomsday Equation, through which someone might manipulate the supercolliders' system codes to begin making Higgs fields. When she is finally told of the Doomsday Equation, Paxton accosts Wyndham: "You and Raj knew there was a 1 in a 1000 scenario where your collider system would destroy the planet and you decided to keep this information to yourselves?" [39]. This accusation and the references to the Higgs field (with the inclusion of the term *vacuum energy* in EVE's name) all point to the screenwriters' potential use of the LHC safety debate as source material for their screenplay.

As the film continues, the Earth's magnetic field and plate tectonics become unstable, and as planes, satellites, and the International Space Station fall from the sky Wyndham and his team survey the remains of Orleans searching for evidence that a Higgs field has been created. Wyndham explains that during the program's early days "alarmists believed that smashing together protons at these kinds of speeds would create some kind of a mini black hole which inevitably leads to the end of the world". When a colleague dismisses that possibility as the delusions of "a bunch of conspiracy theorists", Wyndham has to admit that he and Bashir had discovered that it wasn't impossible, but someone would have to intentionally alter the codes of the system to achieve the Doomsday Equation. The presumed connection to Higgs fields is never clarified, other than the comment by an anonymous team member that "until now we never knew for sure what effects a Higgs field might have on a massive scale" [39].

Bashir is kidnapped by the terrorist who had destroyed the Orleans collider, but frees himself and tries to convince Wyndham that the two remaining colliders are preventing the Higgs field from expanding. Wyndham believes that the colliders are feeding the Higgs field and must be shut down and rebooted. Noble scientist Bashir warns Wyndham that if he shuts down the Geneva facility it will create a black hole, but in the end foolish scientist Wyndham succumbs to Paxton's repeated exhortations that Bashir is a terrorist, and he shuts down the machine. After Wyndham prematurely comforts his son, assuring that they will be okay, the entire Earth is destroyed in a flaming explosion. White letters type across the backdrop of the debris of our planet: "Extinction". It is never explained how the creation of a black hole would cause our planet to explode, but there is much in this film that defies logical explanation. Apart from the troubling stereotypes of both women and Middle-eastern men, the most alarming messages telegraphed by this film are that particle accelerators are dangerous and lack proper security, and, most importantly, that physicists would willingly hide the fact that their experiment posed a risk to our planet as significant as 1 in 1000. *Annihilation Earth* openly and brashly plays to the worst misconceptions and stereotypes of both science and scientists, and it is comforting to know that it is rated at a lowly 3.2 out of 10 by users of the *International Movie Database*.[3]

---

[3] https://www.imdb.com/title/tt1283479/

Discussions of the vacuum energy of the universe also bring up another important but largely mysterious component of the cosmos, dark energy. In 1998 two independent teams of cosmologists, one led by Saul Perlmutter and the other by Adam Reiss and Brian Schmidt, announced the rather unexpected discovery that the expansion of the universe is not slowing down (under the influence of the total gravitational pull of both normal and dark matter) but is actually accelerating, under the influence of some unknown component termed *dark energy*. While the leaders of the two teams shared the 2011 Nobel Prize in Physics, the exact nature of dark energy (now suspected to make up about 68% of the universe) is still a topic of active research. One of the top contenders is the vacuum energy of the universe itself. Given the fact that dark energy makes up significantly more of the universe than dark matter, and in some ways is understood far less, it is not surprising that it figures among the potential destructive powers harnessed by mad scientists in popular culture. *Eve of Destruction* (2013) a two-episode made-for-tv miniseries, recounts the rise and fall of a vacuum "mining" project designed to harvest dark energy as a new source of cheap energy. The Proteus Group's DRIL switch (Dark Radiation Integrated Lattice), the brainchild of scientists Dr. Rachel Reed and her Ph.D. thesis advisor Dr. Karl Cameron, uses a particle accelerator's proton collisions to "dig a well into the vacuum of space" and extract dark energy [40] (Fig. 6.6).

Max Salinger, the mad scientist CEO of Proteus, not only compares Rachel and Karl to famed scientists Richard Feynman and Albert Einstein, but declares them to be the "Adam and Eve of Science" [40]. The "Eve" of the series title should actually be "Eves", as there are three women who play significant roles in the accelerator disaster. When small power fluctuations appear in the system, Rachel is loath to delay the test of her life's work (perhaps even more so due to the government's offer of generous funding if the experiment is a success) despite Karl's concerns. The test proceeds, and when there are fluctuations in the magnetic field, junior scientist Chloe Banks is directed to take two technicians down to the still-running accelerator to make sure there isn't a "magnetic quench incident, like the one they had at CERN". In common vernacular, the event is explained to the head of security as "the magnet overheats, the helium evaporates, everything goes boom. It's very bad" [40].

As expected, a quench does happen; one technician dies, the other is injured, and Chloe and the head of security cover up the quench as a simple magnet malfunction. As one conspiracy begins, a second is uncovered by Karl, a Russian cover-up of a similar experiment that had led to an entire town being removed from all official maps after a 2003 accident. Rachel is still reluctant to accept the dangers of the experiment, even after the destruction

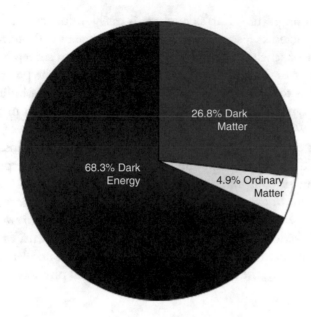

Fig. 6.6 The composition of the universe (User Szczureq, CC BY-SA 3.0, via Wikimedia Commons)

of an entire neighborhood by a massive power spike. When she finally accepts responsibility, Salinger replaces her in the control room with his lover, Chloe. An uncontrollable release of dark energy occurs when the activist group P53 (aided by Karl's rebellious daughter, Ruby) causes a power outage that shuts down the containment apparatus. As an energy vortex begins destroying Denver, Max and Chloe make a quick exit for Mumbai with all of the data (all the while discussing their plans to build another accelerator). With the help of an electric company lineman who, coincidentally, had lost his family in the Russian catastrophe, Rachel and Karl run the experiment again and create a black hole, which consumes all the dark energy and then harmlessly evaporates.

The appearance of three "evil" female characters in one film is interesting, and perhaps it would have been better called "Lilith of Destruction". Rachel is single-minded and driven, but does come around in the end, with the considerable influence of her mentor Karl. Of all of Flicker's female stereotypes, perhaps she best fits the naïve expert, although this description is incomplete at best. The male stereotype of the foolish scientist is perhaps more appropriate, especially juxtaposed against Karl's noble scientist. Chloe is morally corrupt and easily manipulated, especially by her older and powerful lover, Max. She does not act independently, so perhaps she is best described as the evil plotter in training. Ruby, Karl's daughter, is easily influenced and misguided, but is not depicted as a scientist at all. She joins the anti-Proteus group P53 out of a sense of rebellion against her father rather than a deep belief in their mission, and gives them access to the facility, leading to the power outage.

P53's goal is to discover what the secretive Proteus lab is hiding from the public. This type of conspiracy theory is common in much of this particle peril media. Whether directed towards the scientists themselves, their shady financiers, or their political/military allies, the question remains: what aren't we being told about the experiments, their purpose, and, especially, their safety? The air of conspiracy theory is intentional, and speaks directly to a specific segment of the population. However, it is not only particles that are painted as dangerous, as seen in the case of vacuum energy in *Eve of Destruction*. In Chapter 7 we will continue this analysis on a grander scale, as black holes, wormholes, and even entire universes created by particle accelerators threaten our very existence in the hands of pernicious fictional physicists.

References

1. K. Tate, Warp drive & transporters: how 'Star Trek' technology works (info-graphic), Space.com, https://www.space.com/21201-star-trek-technology-explained-infographic.html
2. D. Brown, *Angels & Demons* (Pocket Books, New York, 2000), p. 105
3. D. Brown, *Angels & Demons* (Pocket Books, New York, 2000), p. 108
4. D. Brown, *Angels & Demons* (Pocket Books, New York, 2000)
5. F. Close, *Antimatter* (Oxford University Press, Oxford, 2009), p. 138
6. F. Close, *Antimatter* (Oxford University Press, Oxford, 2009), p. 144
7. L. Zyga, Physicists Scrutinize Antimatter in Angels & Demons, Physics.org, https://phys.org/news/2009-05-physicists-scrutinize-antimatter-angels-demons.html; Julie Steenhuysen, Scientists Debunk 'Angels & Demons' Antimatter, Reuters, https://www.reuters.com/article/us-physics-movie/scientists-debunk-angels-and-demons-antimatter-idUSTRE54I76420090519; Erin McCarthy, Does Angels & Demons Get Antimatter Science Right?, Popular Mechanics, https://www.popularmechanics.com/culture/movies/a4249/4317701/
8. Why does CERN study antimatter?, CERN, https://videos.cern.ch/record/1177161
9. Making antimatter, CERN, https://angelsanddemons.web.cern.ch/antimatter/making-antimatter
10. Can antimatter be used as an energy source?, CERN, https://angelsanddemons.web.cern.ch/faq/antimatter-to-create-energy

11. Does CERN conduct secret research?, CERN, https://angelsanddemons.web.cern.ch/faq/secret-research

12. Is it possible to make a bomb out of antimatter?, CERN, https://angelsanddemons.web.cern.ch/faq/bomb-from-antimatter

13. R. Jayawardhana, *Neutrino Hunters* (Scientific American, New York, 2013), p. 7

14. H. Bethe, R. Peierls, The 'Neutrino'. Nature **133**, 532 (1934)

15. J. Updike, Cosmic Gall, The New Yorker, 36, line 6 (1960)

16. R. Jayawardhana, *Neutrino Hunters* (Scientific American, New York, 2013), p. 125

17. L.M. Krauss, *The Physics of Star Trek*, Rev. edn. (Basic Books, New York, 2007), p. 207

18. A. Strickland, 'Ghost particle' found in Antarctica provides astronomy breakthrough, CNN, https://www.cnn.com/2018/07/12/world/neutrino-blazar-cosmic-ray-discovery/index.html

19. R. Highfield, The quest for dark matter, The Telegraph, https://www.telegraph.co.uk/technology/3340760/The-quest-for-dark-matter.html

20. D. Vebber (script), The Birdbot of Ice-Catraz, Futurama, season 3 (2001)

21. C. Golden, *Shadow of Heaven* (Pocket Books, New York, 2000), p. 255

22. K. Garrett, G. Duda, Dark matter: a primer, https://ned.ipac.caltech.edu/level5/March10/Garrett/paper.pdf

23. L. Randall, *Dark Matter and the Dinosaurs* (Ecco, New York, 2015), pp. 5–7; 55–6

24. K. Vonnegut, *Cat's Cradle* (Dial Press, New York, 2010), p. 50

25. R.L. Jaffe, W. Busza, F. Wilczek, J. Sandweiss, Review of speculative 'disaster scenarios' at RHIC. Rev. Mod. Phys. **72**(4), 1125–1140 (2000)

26. J.-P. Blaizot, J. Iliopoulos, J. Madsen, G.G. Ross, P. Sonderegger, H.-J. Specht, Study of potentially dangerous events during heavy-ion collisions at the LHC: report of the LHC Safety Study Group, CERN, https://cds.cern.ch/record/613175/files/CERN-2003-001.pdf; John Ellis, Gian Giudice, Michelangelo Mangano, Igor Tkachev, & Urs Wiedemann, Review of the safety of LHC collisions, Journal of Physics G, **35**(11), 115004 (2008)

27. A. Brennert (script), Trouble with Harry, Odyssey 5, season 1 (2002)

28. A. Cho, Neutron stars' quark matter not so strange. Science **360**, 697 (2018)

29. P. Donovan (script), Little Blue Planet, Lexx, season 4 (2001)

30. L. Thompson, M. Mazur (dir.), Decay, H2ZZ Productions (2012)

31. D. Overbye, The collider, the particle and a theory about fate. The New York Times, https://www.nytimes.com/2009/10/13/science/space/13lhc.html-NYTimes.com

32. J. Ringo, *Into the Looking Glass* (Baen, Riverdale, 2007), p. 278

33. S. Soderbergh (dir.), Solaris, Twentieth Century Fox (2002)

34. S. Hawking, Foreword, in *Starmus: 50 years of man in space*, ed. by B. May, G. Israelian, (Short Run Press, Exeter, 2014), p. 6

35. R.L. Jaffe, W. Busza, F. Wilczek, J. Sandweiss, Review of speculative 'disaster scenarios' at RHIC. Rev. Mod. Phys. **72**(4), 1130 (2000)
36. J. Leake, Hawking: god particle could destroy universe, The Times, https://www.thetimes.co.uk/article/hawking-god-particle-could-destroy-universe-ds90zfkms3h
37. J. Ellis, CERN Colloquium: The LHC is Safe, CERN, http://cds.cern.ch/record/1120625?ln=en
38. A. Andreassen, W. Frost, M.D. Schwartz, Scale-invariant instantons and the complete lifetime of the standard model. Phys. Rev. D **97**, 056006 (2018)
39. N. Lyon (dir.), Annihilation Earth, Syfy (2009)
40. R. Lieberman (dir.), Eve of Destruction, Reunion Pictures (2013)

7

Specters in Space-Time

7.1 Malicious Intention in the Multiverse

One of the most over the top mad scientists I came across while researching popular media for this book is Dr. Albert Obgrabco from the animated series *The Adventures of the Galaxy Rangers* 1986 episode "Trouble at Texton". Texton is a moon on which the erratic Obgrabco pushes his moon-encircling accelerator to energies way beyond safety limits (in fact beyond the energy of the Big Bang). His goal is to crack open the universe to confirm his hypothesis of parallel universes before his forced retirement. The heroic Galaxy Rangers respond to a distress call from the lab's mainframe computer SAM, but they are unable to stop the evil Obgrabco from realizing his goal. An extraterrestrial from another dimension falls through a crack in space-time and risks its own life to stabilize the fracturing moon long enough for the Galaxy Rangers to escape.

While Obgrabco's hijinks are blatantly (and literally) cartoonish, what happens to scientist Blake Lock in director Andrew Huang's short film *Rift* (2009) is unsettling, to say the least. During a normal family breakfast with his wife and daughter, Lock's eyeglass lenses spontaneously crack, his first clue that this was not to be a normal day. Lock is next seen addressing the press at an accelerator facility news conference, attempting to calm fears about the experiment they are about to conduct. He explains that although "Great scientific discoveries never occur without risk", their plan to create black holes in order to discover parallel universes is safe because the black holes will quickly evaporate. His joking promise that "the sun will come up tomorrow" is met with uneasy silence from the audience, and the fate of the sun soon becomes the

© Springer Nature Switzerland AG 2019
K. Larsen, *Particle Panic!*, Science and Fiction,
https://doi.org/10.1007/978-3-030-12206-5_7

least of their worries [1]. The computer system initially reports a serious error with the experiment, but after a large tremor, the computer announces that the collision has been successful. But what Lock and the others don't immediately understand is that the parallel universes that were created are extremely unstable. Lock shifts between copies of his life that are just a little bit off in some way. In one his wife serves him corn casserole, which he hates, and tells him they have no daughter. In another Lock sees a copy of himself speaking Japanese at the press conference. He finally walks outside of his house and sees everything being sucked into a giant black vortex that has opened up in the sky. A rock cracks his eyeglasses, and he finds himself at his breakfast table once again. He tries to change history by telling his daughter he will be at her soccer game this time (presumably planning to stop the experiment), but walks out the front door only to find his house is floating in space. His house blinks out of existence, followed by Dr. Lock himself.

The term "parallel universe" is thrown around quite liberally in popular culture, and can be used to refer to several very different situations. For example, as noted in Chapter 1, our universe could contain additional dimensions of space that we don't sense in our everyday lives and that the LHC might detect some evidence for. Another use of the term is to describe alternative realities within the Many-worlds Interpretation of quantum mechanics, as described in Chapter 4. This appears to be the basis for *Rift*. Another meaning is to refer to different "bubbles" of space-time evolving parallel to and separately from our own, as predicted in inflationary cosmologies. A collection of different universes of any variety is often termed the *multiverse* [2]. Sometimes the nature of the other universe is purposefully left vague in science fiction. Is it an alternate version of our reality? An inflationary sibling bubble or child universe? Who knows? (Fig. 7.1).

The Voices at CERN is an example of a work that uses higher dimensions of space, specifically an 8th dimension where the minds of the deceased reside. While its existence is only officially discovered when the fictional version of the LHC reaches a collision energy of 14 TeV, it is revealed that very high energy cosmic ray bursts also occasionally allow communication with the 8th dimension, an interesting application of many safety studies' claim that the LHC isn't doing anything that cosmic rays haven't already theoretically been able to do.

MWI alternate realities are a favorite trope for speculative fiction due to the freedom they afford the characters (and author). What would you be doing with your life under slightly different conditions? This is explored in Season 2 of *The Flash*, where the existence of different "Earths" (different realities) is revealed. Characters in the primary reality of the show (termed Earth-1) have

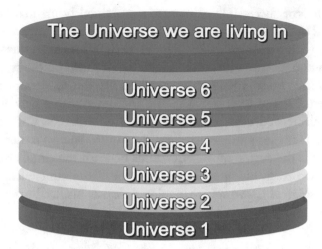

Fig. 7.1 The multiverse as the collection of all universes (regardless of their nature) (User K1234567890y, public domain, via Wikimedia Commons)

doppelgangers on other Earths. For example it is Iris not her father Joe West who is the cop on Earth-2. Dozens of Earths are known (and available for scriptwriters to explore) in *Flash* canon. Portals between these "Earths" were accidentally opened when a planned recreation of the accelerator incident that gave Barry his speed generates a nasty singularity (black hole) as well as ancillary breaches in the fabric of our universe, resulting in evil doppelganger meta-mutants terrorizing the main Earth.

Interdimensional portals unleashing monsters on our world is not a new concept, the gigantic Kaiju monsters of *Pacific Rim* and its sequel being prime examples. Particle accelerators are responsible for such intrusions in *The Black Hole* and *Xtro II*. But perhaps the ultimate "an accelerator caused a monster invasion from parallel worlds" scenario is the direct-to-Netflix film *The Cloverfield Paradox* (2018). The third installment in J.J. Abrams' *Cloverfield* franchise was widely panned, but at least it got online fans (and critics) thinking about the science behind it [3].

In response to a worldwide energy crisis, a revolutionary international space station/particle accelerator is built in orbit around Earth in 2028. The situation back home becomes even direr as each successive test of the accelerator fails, but naysayers warn against its use, especially author Mark Stambler. He claims that every test risks "ripping open the membrane of space-time, smashing together multiple dimensions, shattering reality… everywhere". The result could include "Monsters, demons, beasts from the sea" infecting the past, present and future of all possible universes [4].

The accelerator successfully generates a sustained beam that rises to over 600 TeV and overloads the system. It also throws the crew into another parallel reality in which the station was destroyed (it is suggested due to our universe's accelerator). After much drama and death (both on the station and on their Earth) the two surviving crew members fire the accelerator twice, once to bring them back to their universe and the other with a new algorithm to get the much-needed energy source up and running (again, conveniently ignoring the fact that particle accelerators require energy rather than freely offer it). But as the escape capsule descends beneath the clouds we learn that Stambler was correct, as a ginormous monster of the same alien species featured in the original *Cloverfield* menacingly thrusts his head upward through the clouds.

It is interesting to note that the original unsolicited screenplay (or spec screenplay) for the project was named *The God Particle* (written in 2012 by Oren Uziel), an obvious reference to the Higgs particle. Changes were subsequently made to fit it into the *Cloverfield* franchise [5]. In the original version, the particle accelerator aboard the space station *Dandelion* is actually a secret experimental weapon. This not only ties into general conspiracy fears, but more directly to fictional concerns about Higgs bombs (as in *A Hole in Texas*) and worries that high energy particle accelerator experiments could cause our universe to decay into a new vacuum state (as discussed in Chapter 6). While *The Cloverfield Paradox* openly taps into real concerns about high-energy particle accelerators and alternate realities [6], the satirical website *The Onion* lightens the mood with a 2017 parody of these concerns. Here CERN scientists apologize for the apparent destruction of five parallel universes. It is noted that people might have "experienced momentary vertigo around 9:45 a.m. as a result of several of their alternate identities being wiped from existence" [7].

Marvel comics has also invested in the idea of separate universes intersecting under the right (one might say wrong) conditions. Thor explains in *Thor: The Dark World* that there are nine distinct universes that align every five millennia for a brief event called the Convergence, allowing for travel between these different worlds through space-time rifts. The particle accelerator that the unfortunate technician Tobias of the *Agents of S.H.I.E.L.D.* episode "Repairs" accidentally caused to explode was actually trying to recreate such rifts, causing him to drift, like a ghost, between realities. Similar bridges between "bubble" universes are connected with particle accelerator experiments in John Cramer's *Einstein's Bridge*, allowing for both Earth's invasion by the nefarious Hive and the arrival of aid from the benevolent Makers. The use of the term *bubble universe* in the novel suggests that this work draws from the

Fig. 7.2   Bubbles in soda bring to mind bubble universes (User Spiff, public domain, via Wikimedia Commons)

inflationary cosmology's definition of a multiverse. A similar plotline (and cosmology) is found in John Ringo's *Into the Looking Glass* (Fig. 7.2).

Stories about bridging dimensions and universes have also led to misconceptions thanks in part to careless communication from CERN. Sergio Bertolucci, Director for Research and Scientific Computing at CERN, was understandably ebullient in anticipation of the LHC's return to service in November 2009. *The Register*, a British technology website, reported on a news conference in which Bertolucci enthusiastically described the possibility that the LHC might create "unknown unknowns" including a temporary doorway to another dimension. In particular, Bertolucci is quoted as offering that "Out of this door might come something, or we might send something through it", further explaining in a follow up communication with the website that although such a doorway could only be opened for a miniscule fraction of a second, "during that infinitesimal amount of time we would be able to peer into this open door, either by getting something out of it or sending something into it". While Bertolucci adds that such a connection to another dimension would pose "no risk to the stability of our world", the damage had already been done in terms of the online conspiracy community [8]. The *Register*'s story was widely reported on conspiracy websites, including *Rapture Ready*, where Matt Ward wrote that he found statements by Bertolucci and other CERN scientists "Stunning in their simplicity, in their childlike abandonment of consideration for risk and stunning in their arrogance and hubris" [9]. While these charges are probably simply motivated by a lack of understanding of the basic physics, my hope is that scientists can pause for a moment and try to see the argument from the side of the anxious nonscientist. Recall that physicists also voiced considerable excitement about the possibility of the LHC being a "black hole factory", something the average person might find difficult to picture as positive, as we next explore [10].

7.2   Black Holes Are Bad News

Leave it to science fiction writers to find a "practical" use for tiny black holes, for example as propulsion systems in Dan Simmons' novel *Hyperion* and in Romulan warbirds in *Star Trek: The Next Generation*. But artificially created black holes are more often seen as potential liabilities. In his afterword to *Earth*, David Brin explains "nothing spices up a novel like a monster threatening to gobble up the world" [11]. In J.J. Abrams' *Star Trek* reboot (2009), Romulans drill a hole into the core of the planet Vulcan and inject it with a "red matter" bomb that creates a planet-devouring black hole. The children's live-action series *Lab Rats* episode "Back from the Future" (2012) finds superhero siblings Adam, Bree, and Chase risking their lives to shut down an out-of-control particle collider before it can create a black hole that will "implode the entire planet" [12]. There is another common scenario for mini black holes in science fiction, a gravitational death by a thousand paper cuts.

Recall from Chapter 3 that a Hawking black hole that is moving too slowly to escape Earth's gravity could become embedded within our planet, oscillating back and forth through the center, slicing small tubes of damage through the rock with each pass. The issue is whether the black hole is ultimately gaining mass through directly colliding with or gravitationally pulling into itself Earth's individual atoms, or evaporating its way towards a final explosion through the Hawking mechanism. Not surprisingly, such black holes appear at the heart of numerous science fiction tales. The microscopic black hole of the novel *The Krone Experiment* does surgically precise (yet deadly) damage around the world as it pierces in and out of the Earth's surface. From eviscerating a sheep in the foothills of the Atlas Mountains to exploding the fuel tank of a Russian carrier, releasing deadly chlorine gas from a punctured chemical tank in Nagasaki to blowing a hole in a South African politician's aorta and causing a mine collapse in China, a tiny black hole oscillating back and forth through our planet can certainly ruin your day.

Destroying our own planet through experimental carelessness is one thing, but what about when an "oops" leads to the death of a celestial neighbor such as the Moon (as in the case of Paul McAuley's short story "How We Lost the Moon") or Mars? Larry Niven's Hugo Award winning short story "The Hole Man" (first published in *Analog* magazine in January 1974) is Andrew Lear's account of his part in the future destruction of Mars. The crew of the ship *Percival Lowell* finds an alien base on the Red Planet, abandoned except for a gravity wave communicator intentionally left running. Lear surmises that there is a microscopic black hole inside that has been trapped there for about 600,000 years. While playing around with the dials, Lear accidentally (or not—it's debatable) shuts off the containment field, causing the black hole to fall through the floor and through the body of the bullying Commander Childrey, who bleeds to death. Lear calculates that the black hole (whose source is never revealed) will take well beyond his lifetime to completely devour Mars. Were the extraterrestrials guilty of creating the object in the first place, or had they merely trapped it there in order to protect Mars from the very fate Lear had doomed it to? Astronaut Teresa Tikhana of David Brin's novel *Earth* reflects that perhaps making "world-wrecking black holes" is easy for technologically advanced species; therefore the reason why we haven't been contacted by extraterrestrials is that at some point each civilization creates "unstoppable singularities, and gets sucked down the throat of its own, self-made demon" [13].

If Hawking is right, there are microscopic black holes throughout the universe. What's to stop one from just showing up on our doorstep one day? This possibility is explored in *The Doomsday Effect* by Thomas Wren (a penname of Thomas Thurston Thomas), briefly mentioned in Chapter 3. Winner of the 1987 Compton Crook Award, the novel begins with a microscopic black hole piercing through a plane flying over the San Andreas Fault. It nearly kills Pinocchio robot company chief engineer Alex Kornilov and the Executive Vice President for Marketing Grace Porter and sets off an 8.8 magnitude earthquake. Geophysicist Ariel Ceram convinces Pinocchio CEO Steve Cocci that the microscopic black hole they dub Hawking-1 not only set off the tectonics plates and gave Grace and Alex severe cases of radiation sickness (due to the black hole's tiny accretion disk), but it will slowly eat the planet over millions of years. Cocci and Ceram initially argue about what they should do about the problem (Cocci interested in profit and Ceram the science), but they team up after more careful calculations demonstrate that the Earth has less than seven years left. On the other hand, the Defense Department wants to turn Hawking X-1 into a weapon by changing its trajectory to intersect enemy targets. Cocci and Ceram's solution to dealing with both military conspiracies and public conspiracy theories is to go public, so public that the average person will be fascinated rather than fearful. The hope is that the public will follow the story so closely that the government will not be able to coopt the black hole for military purposes, giving scientists time to work out a plan to dispose of the nasty little pest (a rather ingenious manipulation of the popularization of science).

If you were a mad scientist, you might intentionally feed your pet black hole to make it bigger, as in Larry Niven's Hugo Award winning novelette "The Borderland of Sol" (first published in *Analog* magazine in January 1975). Dr. Julian Forward searches for the 1908 Tunguska object described in Chapter 3, believing it to potentially be a black hole that escaped Earth's grasp. His funding is cut just as he discovers an unrelated tiny black hole, which he artificially enlarges to a mass comparable to that of an asteroid. Containing the object in his asteroid base laboratory, Forward procures resources by stealing them from unsuspecting spaceships, using his pet black hole to devour the evidence of his crimes. As one might suspect, the containment mechanism fails, and the mad scientist falls victim to his own creation.

It is important to remember that there are competing effects at work when discussing microscopic black holes. Eva Sondstrom explains in a long expositional scene in *The Void* that Abernathy's tiny manufactured black hole could sink into the Earth if it were not contained (making comparisons to the film *The China Syndrome* for the sake of her less-than-genius engineer boyfriend) and find itself oscillating back and forth as it consumed more material, growing with each pass. The other possibility she offers is that it could explode, a reference to the Hawking mechanism. As explained in Chapter 3, being in close proximity to a black hole that is radiating via the Hawking mechanism isn't exactly healthy for any human being, but it is certainly a better option for the planet itself. In fact, as described in Chapter 5, the probability that any microscopic black hole created by the LHC would almost instantaneously decay via the Hawking mechanism is central to many safety reports on the subject.

Recall from Chapter 5 that Steven Giddings had been contacted by a journalist inquiring what would happen to tiny black holes potentially created at the LHC if the Hawking mechanism isn't real. In a follow-up paper Giddings offers that "we certainly believe[1] that Hawking's calculations are correct, if not to the last detail, and furthermore on general quantum grounds black holes

---

[1] There's that pesky B-word again, being used in a scientific context, when the word "expect" would have been more accurate and neutral (in terms of not confusing it with an article of faith).

should decay". Giddings concludes by suggesting that "further assurances are welcome" [14].

However, some of the "assurances" that the physics community afterwards received are couched in language that could cause alarm in the minds of non-specialists, especially those who are predisposed to seeing danger around every corner where the LHC is concerned. Recall that Hawking had discovered this predicted mechanism by applying quantum mechanics to our understanding of black holes. But since we do not currently have a Theory of Everything that explains exactly how one would wed quantum mechanics to gravity, it is possible that Hawking's work only holds up under certain conditions. This is precisely what William G. Unruh and Ralf Schützhold explored in 2004. By imposing several general and reasonable assumptions they found that Hawking's predictions hold independent of the details of a TOE, although the authors offer examples of situations that did show what they term "deviations from Hawking's result". They conclude that "whether real black holes emit Hawking radiation or not remains an open question and gives non-trivial information about Planckian physics", meaning that the discovery of Hawking radiation would give us valuable information about a TOE [15].

Walter Wagner, the author of the *Scientific American* letter to the editor concerning RHIC and a plaintiff in anti-LHC lawsuits described in Chapter 5, used Unruh and Schützhold's conclusion as evidence that the LHC is potentially unsafe. Asked to respond to Wagner's claims about the meaning of his conclusion, Unruh explains in an interview that Wagner misunderstands the result of the paper, and that for black holes to not evaporate the universe would "really, really have to be weird" [16]. Looking at the scientific paper in hindsight, it is possible to see how a nonscientist could have easily misinterpreted the conclusion if they did not have the necessary background to digest the main body of this rather dense work. This episode illustrates Giddings' earlier point that physicists should take note that non-specialists are reading online archives of scientific papers. For anyone still feeling a tad anxious at Unruh's conclusion, his abstract of a 2007 conference talk states of the Hawking mechanism that "the effect is almost certainly right" [17]. It is important to note that until scientists actually observe a microscopic black hole and watch it evaporate, we can't be 100% certain that this is the way the universe works.

Giddings also took part in another public discussion of black hole evaporation, centered on a preprint released a few months before the initial startup of the LHC. In response to continued safety concerns about the possibility of creating black holes that would grow and not evaporate if unleashed inside the Earth (as well as his own desire to improve the scientific basis for such analysis) Giddings and Michelangelo Mangano explored what the large-scale effects of a stable black hole would be [18]. They found that there could be no macroscopic effects to our planet on time scales shorter than the life of the Sun because any black holes created by cosmic rays would have caused visible damage to (or destroyed) white dwarfs and neutron stars before now. They conclude that "there is no risk of any significance whatsoever from such black holes" [19].

Astrophysicist Rainer Plaga issued a dissenting arXiv preprint shortly afterwards, suggesting that potential risk remains for certain scenarios [20], claims that were tersely picked apart item by item by Giddings and Mangano in a later preprint [21]. While Plaga's arguments have been largely dismissed by the scientific community, they have been adopted by LHC critics as proof that the community of particle physics experts is either hiding the risk or refuses to acknowledge it [22]. Plaga's public debate with Giddings and Magnano is only part of a wider scientific discussion on possible black hole production at the LHC that occurred leading up to the machine's start-up in 2008. Papers by a number of scientists also added their own analysis to the choir of voices singing the overall safety of the LHC during this time [23]. The volume of the discussion (both in terms of size and intensity) could lead suspicious individuals to question whether or not physicists have a good handle on the situation, or are simply protesting too much. While we should certainly not curtail the open exchange of scientific information and debate, scientists need to be aware of the effects that such debates might have on the general public, and not act surprised when such a flurry of scientific activity leads to further conspiracy theories. Honest, clearly written blog posts (and other uses of social media) by the scientists involved can counter some of this fearmongering, by openly inviting the public to be informed spectators in the scientific process.

An interesting example involves theoretical cosmologist Ian Moss, who exhibited a refreshing acknowledgment of how his calculations could unintentionally feed public fears. In a 2015 paper, Moss, Philipp Burda and Ruth Gregory describe a model in which a microscopic black hole, either a primordial black hole or one potentially produced by the LHC, could act as a seed to trigger the transition to a true vacuum [24]. Moss notes in an interview for *Science* that since primordial black holes (if they exist) haven't done so already, and high-energy cosmic rays haven't created transition-triggering black holes, there is no need to worry. Instead, the lack of a phase transition suggests that there is more to learn about our universe. More relevant to our immediate discussion is Moss's admission that he was concerned that the paper's results could be misconstrued by nonscientists, and voiced fear that John Ellis would accuse him of "scaremongering" [25]. When asked, Ellis did admit that the paper had some potential to "whip up unfounded fears about the safety of the LHC" again, but that he was not going to "lose sleep over it" [25]. As Ellis notes, you certainly can't tell physicists how to present their research. What we do need is honest public discussion, as happened in this case. Moss is probably sleeping better at night himself knowing that there appears to be no widespread use of his work by LHC critics as of now, despite articles by science writers (who should know better) describing his work with leading titles like "Could Black Holes Destroy the Universe: Before you say 'no way,' hear the latest research out" [26]. In contrast, the headline from the *International Business Times* website simply states "Mini black holes from LHC are not going to bring about the collapse of the universe", which is much closer to what the public really wants (and needs) to know [27].

This is a fitting opportunity to revisit the CERN *Angels & Demons: The Science Behind the Story* webpages, in particular their answer to "Does CERN create black holes?" [28]. The first paragraph is very clear ("The creation of black holes at the Large Hadron Collider is very unlikely"), honestly worded ("However, some theories suggest the formation of tiny 'quantum' black holes may be possible"), and meant to turn potential fear into wonder ("The observation of such an event would be thrilling in terms of our understanding of the Universe"). The punchline of this introduction is strong, calling the creation of tiny black holes at CERN "perfectly safe". The page makes a clear distinction between large black holes formed in the deaths of massive stars and microscopic black holes before reiterating that CERN experiments would still be safe if small black holes were ever to be made there because cosmic rays are higher energy cousins to CERN experiments. A link to the LHC safety reports is also included for interested readers. Overall, the page employs a number of best practices in debunking misconceptions and is an effective resource for the general public.

7.3 Troubles in Space-Time Travel

The Japanese anime tv series *Darker than Black: Gemini of the Meteor* (2007–10) follows the aftermath of mysterious spatial anomalies called gates appearing in South America and Japan. Out of these gates emerge thousands of humanoids with superpowers, called contractors. The existence of the contractors is a secret protected by the various shadowy organizations that utilize them as assassins. One of these, the Syndicate, attempts to destroy all the contractors using anti-gate particles emitted by the sun. The plot is to accelerate these using the Saturn Ring particle accelerator and smash them into the

Toyko gate, destroying it and closing the wormhole. The fact that a similar plan had failed five years before is conveniently ignored (as it was in *The Void, Dark Storm*, and numerous other works).

A different kind of gate is central to the short-lived tv series *Terra Nova* (2011). Here scientists at Fermilab discover rifts in space-time that allow time travel into the past. After 22 years of testing and technological advances Hope Plaza is constructed, where humans travel through such a rift 85 million years into the past in order to save our species from overpopulation and pollution. In order to avoid possible paradoxes caused by changing history, it is explained that the "past" of the Terra Nova colony is in a different, alternate timeline, so unfortunately nothing can save the present Earth of the series. Similarly, the desperate time travel experiments in Gregory Benford's novel *Timescape* serve to warn the scientists of 1962–3 of the extreme environmental catastrophes to come. However, this intervention forces a split into an alternate timeline (one in which John F. Kennedy is not assassinated, for example). John Carpenter admits to borrowing from his friend Benford's novel in writing his science fiction-horror film *Prince of Darkness* in which a physics professor and his graduate students experience a warning from the future every time they fall asleep [29].

Some science fiction authors appeal to specific types of space-time defects in an attempt to be as faithful to the science as possible. In 1985, astronomer and science popularizer Carl Sagan reached out to his friend Caltech theoretical physicist Kip Thorne for help with a plot point in the science fiction novel Sagan was then writing. Called *Contact*, the novel's plotline required the heroine, radio astronomer Eleanor Arroway, to travel light years away from Earth in just an hour. Sagan was going to explain the faster-than-light travel by having Eleanor pop into a black hole near Earth and pop out again near another star. Thorne informed his friend that this was not physically possible by the known laws of physics, instead suggesting a wormhole as a shortcut [30]. A wormhole connects two distant locations in space (called the mouths) via a "throat" that is much shorter than the distance one would have to travel through normal space between these two points. The laws of physics also permit time travel using wormholes (under the right conditions), allowing a traveler to arrive home before he or she even left. Work by Thorne and his collaborators has demonstrated that, while it is hypothetically possible for wormholes to be created under the right conditions, they would be very, very difficult to stabilize and hold open [30] (Fig. 7.3).

Despite the laws of physics conspiring against creating a wormhole rapid transit system any time in the near future (if ever), the possibility of getting around the frustrating limitations of the speed of light (or even traveling

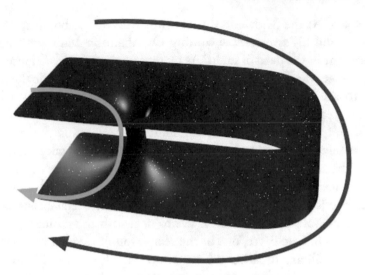

Fig. 7.3   A wormhole as a short-cut in space (User Panzi, CC-BY-SA-3.0, via Wikimedia Commons)

between separate universes) is very appealing to science fiction writers. Whether they are called wormholes, rifts, time tunnels, breaches, gates, Einstein-Rosen bridges, or even windows between worlds (as in Philip Pullman's *His Dark Materials* series), fictional short cuts through space and time are one of the staples of science fiction. But how does one create such a shortcut? A particle accelerator seems a likely scenario, no? John Ringo seemed to think so when he wrote *Into the Looking Glass*. In this novel the explosion of a particle accelerator creates Higgs particles that somehow generate wormholes connecting our universe to that of the pernicious Dreen. Throughout the novel humans travel through wormholes from one universe to another as they fight the Dreen and their accomplices, destroying space-time gateways that open up to hostile universes. The title of John Cramer's novel *Einstein's Bridge* refers to a wormhole that is created to help two physicists travel back in time to prevent the creation of the SSC (in order to save Earth from an alien attack precipitated by the SSC's high-energy collisions). In an interesting counterpoint to these two novels, in Patrick Lee's *Deep Sky* it is revealed that immortal humans living in another star system in the future use an ancient alien wormhole to connect with their past, in particular the inaugural firing of a particle accelerator on 1978 Earth.

If your fictional particle accelerator connects with a dangerous wormhole, why not also add a black hole and throw in an electricity-eating monster for good measure? This is the basis for the film *The Black Hole*, briefly introduced in Chapter 5. At the Midwestern Quantum Research Laboratory (MQRL), Dr. Hauser and his associates, including Dr. Shannon Muir, are conducting an experiment in the dead of night. As he readies the controls, Hauser smugly notes, "let's see what God has in store for us tonight" [31]. While the reference may be to the Higgs particle, it can certainly be interpreted as a warning against the stereotypical urge of scientists to play God. Following an explosion and radiation leak, Hauser dismisses Shannon's suggestion that they follow the proper safety protocol and report the incident, instead taking a male associate into the accelerator tunnels. There they discover that the experiment has not only unexpectedly spawned a far-from microscopic black hole, but an electromagnetic monster. The creature kills Hauser, the associate is sucked into the black hole, and Shannon is left to deal with the military response to the accident. Dr. Eric Bryce, one of the team's founding members, is brought back by the military to assess the situation, and finds that his former colleagues had decided to throw caution to the wind by voting to accelerate their timeline under the pressure of competition from a Chinese laboratory. Eric, in his role as the noble lone voice, accosts Shannon and the team for entrusting the fate of the planet to a simple majority vote. One would hope that the viewer does not believe that CERN makes decisions in this cavalier manner.

Eric realizes that the black hole grows as the monster takes in more electric energy, comparing it to "particles connected over long distances on the quantum level" (invoking the phenomenon of quantum entanglement discussed in Chapter 4) [31]. Eric reasons that the creature must be using wormholes to travel across the universe, and that Hauser's experiment created a weak spot in the space-time fabric of the universe that allowed the black hole—said to be one of the mouths of the wormhole despite Kip Thorne's research to the contrary—to open up in the lab. As the military prepares to drop a small nuclear missile on St. Louis, Eric and Shannon manage to lure the monster into the black hole by using a souped-up electric generator truck, sending both entities somewhere else in space and time, and thus saving the world. *The Black Hole* is not to be commended for either its scientific liberties or its portrayal of scientists. Not only are the particle physicists painted as irresponsible in violating established safety protocols, but Shannon is largely depicted as weak and ineffective. She is the perfect example of Eva Flicker's daughter/assistant stereotype, as she has no meaningful role except in relation to her former lover Eric and her male colleagues.

As noted in previous chapters, *The Flash* also draws upon wormholes and breaches between realities but includes somewhat more positive (although certainly still flawed) portrayals of women scientists. At the end of Season 1 Barry and his colleagues create a wormhole using the S.T.A.R. Labs particle accelerator in order to allow travel to the past and future. But due to an unexpected time paradox (the evil Eobard's forefather commits suicide long before Eobard is born), the wormhole turns into a gigantic black hole over Central City. While Barry and his meta-human friends manage to dissipate the black hole, over 50 breaches open up between their universe and the universe of the evil speedster Zoom, creating convenient sources for dramatic storylines. Dr. Tina McGee, a "brilliant but egocentric physicist", also develops a tachyon device that allows a "speedster" like the Flash to run even faster [32]. Tachyons (hypothetical particles that can only travel faster than light) are also referenced in both the novel and television adaptation of *Flashforward*. In the novel Theo Procopides is Director of the "Tachyon-Tardyon Collider" at CERN in 2030 (at least according to the flashforward visions) [33]. In the television series one mystery surrounding the global blackout conspiracy is the so-called "Tachyon Constant", described as a "theoretical number that physicists have been trying to crack for centuries. Sort of a holy grail thing" that allows the time of the next global blackout to be computed [34]. Tachyons are also used to send information from the future into the past in *Prince of Darkness* and *Timescape*.

While the previous scenarios that link particle accelerators and the creation of wormhole-like structures are clearly science fiction, Russian mathematical physicists Irina Ya. Aref'eva and Igor V. Volovich have suggested that the LHC has about the same probability of creating so-called traversable wormholes (ones that remain stable enough for travel between the two mouths) as it does of creating black holes [35]. Since a well-known limitation of theoretical time machines is that you can only travel back in time to the moment that the machine was turned on and not before (a point that is conveniently ignored in science fiction), time travelers using LHC created wormholes could not travel back in time to before the initial firing of the machine. In an article by science writer Roger Highfield in the British newspaper *The Telegraph*, Oxford physicist David Deutsch calls the possibility "speculative in the extreme but not cranky" [36]. University of Manchester particle physicist and well-known popularizer of science Brian Cox jokingly offers to "personally eat the hat I was given for my first birthday before I received it" if wormholes are found at CERN [36]. The idea of wormholes being created at CERN is therefore not considered a good bet, either in the scientific or gastronomical sense.

7.4   Bringing Up Baby (Universes)

The Simpsons 2012 Halloween special "Treehouse of Horror 23" not only pokes fun at the then upcoming December 21, 2012 Mayan calendar apocalypse, but the possible creation of black holes at particle accelerators. But this particular black hole has a surprise up its sleeve (or rather, in its pocket). The inauguration of the Springfield Subatomic Supercollider—"which will either answer certain obscure questions of subatomic physics or destroy the universe"—is heralded with much hoopla until nerdy scientist Frank throws a giant ON switch and absolutely nothing appears to happen [37]. Overlooked by the crowd, a black hole the size of a baseball bursts out of the collider ring and begins slurping up matter. Lisa finds the black hole and brings it home like a stray cat to keep it out of trouble after she sees it suck up a child. Her family members amuse themselves by using the object as a garbage disposal (against Lisa's warnings), causing it to grow. Her father even goes so far as to start a waste removal business, which causes the black hole to become so large that it not only eats their house, but sucks in the entire town, except for the Simpson family infant Maggie, whose pacifier plugs up the opening to the black hole. In the interior of the black hole, the citizens of Springfield find themselves on an alien planet whose residents refer to them as "trans-dimensional visitors" and thank them for the treasures that have appeared in their universe via the black hole [37].

Since the event horizon of a black hole is exactly that, a boundary between events we can observe (the exterior) and those we cannot (the interior), and since, in our current understanding of the laws of physics, the matter that formed the black hole is said to be compacted down to a mathematical point called a singularity, how in the world can it make sense to talk about an entire universe inside a black hole? The key point is that the other universe is not "inside" the black hole, but, rather, what we see as a black hole in our universe is part of an umbilical cord of space-time connecting our universe to a separate child universe that was spawned from our space-time. As science fiction as this sounds, the possibility is based on sound (if not slightly speculative) science. According to the inflationary extension to the Big Bang model, our universe shares a common birth with many other bubbles of space-time that have evolved independently as separate universes since the earliest moments of creation. The sum of all these inflationary bubbles is one example of a multiverse (Fig. 7.4).

In a 1987 article in *Physics Letters* Edward Farhi and Alan Guth, the father of the inflationary cosmology, pondered the possibility of intentionally creating an inflationary bubble universe, a so-called child universe, in an experi-

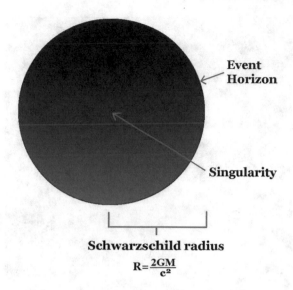

Fig. 7.4 Structure of a simple black hole (User Tetra quark, CC BY-SA 4.0, via Wikimedia Commons)

ment. They found that the initial singularity needed to start the universe growing (the equivalent to the singularity of our own Big Bang) was apparently an "insurmountable obstacle to the creation of an inflationary universe in the laboratory" [38]. However, a 1990 paper by Farhi, Guth and Jemal Guven found that the same quantum mechanical process that could one day turn our current false vacuum state into a true vacuum might also allow a microscopic amount of false vacuum (not dense enough to be a singularity, but still not something you can just whip up in the lab) to grow to form a new, independent universe. Note that the child or pocket universe is not expanding to fill up our universe, but is expanding into a new "space" separate from our own [39]. The singularity seen in our universe is temporarily connected to a singularity seen on the other side (in the child universe); both will evaporate via the Hawking mechanism, eventually breaking the tie between the universes like the cutting of an umbilical cord. Other physicists have found that this thought experiment does not violate the laws of physics in principle (as we currently understand them), but would be very, very difficult to accomplish [40] (Fig. 7.5).

Importantly, it is not impossible, and just the idea that it *might* be possible is certainly good enough for science fiction writers. How would one generate the necessary initial state to form a child universe? A particle accelerator, of course, especially in the case where Guth's initial paper (requiring a singularity or black hole) is the inspiration. For example, Garfield Reeves-Stevens'

Fig. 7.5 Alan Guth (Betsy Devine, CC BY-SA 3.0, via Wikimedia Commons)

scientists actively discuss Guth and his colleagues' work on child universes in the novel *Dark Matter* after mad scientist Cross claims that he can create singularities in the lab. His colleagues worry that these experiments will destroy our universe, but instead only succeed in trashing the laboratory. Despite this setback, government agent Paine is eager to duplicate the experiment for its potential military uses. The novel ends with the now hopelessly insane Cross using his ability to control quantum fields to create a singularity and a child universe into which his and his lover Charis Neale's consciousnesses escape before the government's nuclear weapon drops on them. David Brin's *Earth* concludes with a short story named "Ambiguity" set a few years after the events in the novel. Alex Lustig, who had created and accidentally released Hawking black holes in the main novel, is now experimenting with the practical uses of singularities in a lab on an asteroid beyond Mars. He creates a singularity, but by feeding it virtual pairs from the vacuum, he grows it to planetary mass, and then it disappears, becoming its own child universe.

Former CERN scientist Stan Goldman wonders if our own universe was created in the same way, as someone's little experiment.

The possibility of creating a child universe does bring up such philosophical thoughts, and even parental feelings of responsibility, especially in the case of Gregory Benford's *Cosm*. As described in Chapter 2, after the destruction of her particle detector at the RHIC accelerator, Alicia Butterworth and her postdoc steal the mysterious sphere that had been created and set it up in her University of California—Irvine laboratory. As it grows from the size of a baseball to that of a small bowling ball the sphere utterly confounds her, and she seeks out the advice of Caltech assistant professor of astrophysics Max Jalon. It is Max who discovers that the sphere is a child universe evolving at an extremely accelerated rate of speed (as seen from our side), as would be necessary for the purposes of the novel. Alan Guth's research papers on the topic are mentioned by name and sufficient explanation is given to satisfy the demanding reader. Despite the dangers, Alicia's experiment is repeated at RHIC, resulting in the destruction of the accelerator and the creation of a much larger child universe that has different laws of physics from Alicia's cosmological offspring.

Like any good parent trying to protect their child, Alicia and her colleagues steal the sphere from the observatory where it had been quarantined in order to protect it (and her) from both religious fanatics who have tried to kidnap her and the government officials who want to control (and potentially exploit) her experiment. While hiding in the desert, Alicia, Max and others observe the last stages of the child universe's evolution. The stars and galaxies of Cosm die as extraterrestrial civilizations attempt in vain to rage against the literal dying of the light. Suddenly the mouth of the umbilical cord wormhole on the Cosm side is destroyed by a black hole at the center of a galaxy, causing the sphere to explode, conveniently just as the federal agents find Alicia and her co-conspirators. Although the ability to clearly see into the child universe through a wormhole and watch its evolution is a necessary plot device (because staring at a small black hole doing nothing but emitting Hawking radiation would make for a particularly dull novel), the underlying concept of child universes is rooted in peer-reviewed science.

While the novel is notable for its hard science fiction plot, the stereotypical portrayal of Alicia Butterworth's relationships with men is distressing. At their first meeting in his office Max assumes she is a student, despite the fact that they have the same academic rank (unfortunately not an uncommon assumption made of women in physics back in the 1990s, as I can attest). As the head of her own research lab with a male postdoc and male Ph.D. student under her direction, Alicia is clearly in a position of responsibility. However, her authority is openly challenged by her graduate student, and when her voice (and confidence) waver her male post doc has to step in to defend her. When the existence of Cosm becomes public, a *National Enquirer* headline calls her "Girl who makes Galaxies" the use of "Girl" demeaning despite the fact that it is probably simply meant to be alliterative [41]. While Alicia is able to use her intelligence to pop open the trunk of a car from the inside and escape when she is kidnapped by religious extremists, Max saves her from a potential date-rape situation. Predictably, Alice and Max become romantically involved, and in what is perhaps the most egregious example of institutionalized sexism in the novel, the work ends with a series of newspaper headlines, including the social announcement in a local newspaper that "Dr. Thomas Butterworth of Palo Alto proudly announces the engagement of his daughter, Alicia, to Dr. Max Jalon" [42]. The unfortunate absence of the title "Dr." in reference to Alicia returns us to a question posed in Chapter 2 as to which stereotype of female scientists best fits Alicia Butterworth. In the end it appears that we are to recognize her as an old maid type, who finally shifts from being married to her work to a literal (and socially expected) marriage to a strong male figure who it is (perhaps not so subtly) suggested is a "better" scientist than she is. It is important to note that he is a theorist and she is an experimentalist, an important division within the hierarchy of the physics community, with the theorists seen by some to have "more status" [43].

While the damage caused to particle accelerators in *Cosm* plays into fears of uncontrollable (and unexpected) consequences of high-energy experiments, it is a refreshing twist to have an accelerator as a creator of a universe rather than a destroyer of one (despite the sexist typecasting of a female scientist as the "mother" rather than a male scientist as father). As we have seen, even marginally positive messages are too few and far between in science fiction featuring particle physics experiments. Our final chapter will circle back to a discussion of science communication, especially examples of how the particle physics community has alternately been its own best advocate and worst enemy in terms of trying to highlight the joys of discovery and assuage unfounded fears.

References

1. A. Huang (dir.), Rift, Intelligent Life Productions (2009)
2. M. Tegmark, Parallel Universes, Sci. Am. **288**, 40–51 (2003)
3. P. Hess, The Real Science of the God Particle in Netflix's *The Cloverfield Paradox*, Inverse, https://www.inverse.com/article/40935-cloverfield-paradox-higgs-boson-god-particle; V. Mukunth, *The Cloverfield Paradox* Is a Textbook Case of How

Not to Use Physics in a Movie, The Wire, https://thewire.in/film/the-cloverfield-paradox-review

4. J. Onah (dir.), *The Cloverfield Paradox*, Paramount Pictures (2018)

5. R. Lambie, *The Cloverfield Paradox*: How God Particle Changed, Den of Geek, http://www.denofgeek.com/us/movies/cloverfield/270788/the-cloverfield-paradox-how-god-particle-changed

6. S. Gallagher, *The Cloverfield Paradox*: What Does The Ending Really Mean?, WhatCulture, http://whatculture.com/film/the-cloverfield-paradox-what-does-the-ending-really-mean

7. CERN Researchers Apologize For Destruction Of 5 Parallel Universes In Recent Experiment, The Onion, https://www.theonion.com/cern-researchers-apologize-for-destruction-of-5-paralle-1819579830

8. L. Page, Attack of the Hyperdimensional Juggernaut-Men, The Register, https://www.theregister.co.uk/2009/11/06/lhc_dimensional_portals/

9. M. Ward, Cern [sic], the Large Hadron Collider and Bible Prophecy, Prophecy Update, http://www.prophecyupdate.com/cern-the-large-hadron-collider-and-bible-prophecy.html

10. S.B. Giddings, S. Thomas, High Energy Colliders as Black Hole Factories: The End of Short Distance Physics, ArXiv, https://arxiv.org/abs/hep-ph/0106219

11. D. Brin, *Earth* (Bantam Books, New York, 1994), p. 660

12. R. Rappaport (script), Back from the Future, Lab Rats, season 1 (2012)

13. D. Brin, *Earth* (Bantam Books, New York, 1994), p. 427

14. S.B. Giddings, Black Hole Production in TeV-Scale gravity, and the Future of High Energy Physics, ArXiv, https://arxiv.org/abs/hep-ph/0110127

15. W.G. Unruh, R. Schützhold, On the Universality of the Hawking Effect, ArXiv, https://arxiv.org/abs/gr-qc/0408009

16. D. Overbye, Asking a Judge to Save the World, and Maybe a Whole Lot More, The New York Times, https://www.nytimes.com/2008/03/29/science/29collider.html

17. B. Unruh, Where Do the Particles Come From?, Effective Models of Quantum Gravity Conference, https://www.perimeterinstitute.ca/conferences/effective-models-quantum-gravity

18. E. Vance, Outsmarting the CERNageddon, Nautilus, http://nautil.us/issue/4/thc-unlikely/outsmarting-the-cernageddon

19. S.B. Giddings, M.L. Mangano, Astrophysical Implications of Hypothetical Stable TeV-scale Black Holes, ArXiv, p. 1, https://arXiv.org/pdf/0806.3381.pdf CERN-PH-TH/2008-025

20. R. Plaga, On the Potential Catastrophic Risk from Metastable Quantum-black Holes Produced at Particle Colliders, ArXiv, https://arXiv.org/abs/0808.1415v1

21. S.B. Giddings, M.L. Mangano, Comments on Claimed Risk from Metastable Black Holes, ArXiv, https://arXiv.org/pdf/0808.4087.pdf

22. E. Penrose, A Critical Review of Safety Papers Concerning Black Holes at the LHC, Risk Evaluation Forum, http://www.risk-evaluation-forum.org/LHC_safety.pdf

23. B. Koch, M. Bleicher, H. Stöcker, Exclusion of Black Hole Disaster Scenarios at the LHC, ArXiv, https://arxiv.org/abs/0807.3349; M.E. Peskin, The End of the World at the Large Hadron Collider? Physics **1**, 14 (2008); R. Casadio, S. Fabi, B. Harms, Possibility of Catastrophic Black Hole Growth in the Warped Brane-world Scenario at the LHC, ArXiv, https://arxiv.org/abs/0901.2948

24. P. Burda, R. Gregory, I.G. Moss, Vacuum Metastability with Black Holes, ArXiv, https://arXiv.org/pdf/1503.07331.pdf

25. A. Cho, Tiny Black Holes Could Trigger Collapse of Universe—Except That They Don't, Science, http://www.sciencemag.org/news/2015/08/tiny-black-holes-could-trigger-collapse-universe-except-they-dont

26. S. Hossenfelder, Could Black Holes Destroy the Universe, Starts with a Bang, https://medium.com/starts-with-a-bang/could-black-holes-destroy-the-universe-de8a3135856f

27. H. Osborne, Mini Black Holes from LHC Are Not Going to Bring About the Collapse of the Universe, International Business Times, https://www.ibtimes.co.uk/mini-black-holes-lhc-are-not-going-bring-about-collapse-universe-1514203

28. Does CERN Create Black Holes?, CERN, https://angelsanddemons.web.cern.ch/faq/black-hole

29. J. Beahm, M. Felsher, C. MacMillan, Sympathy for the Devil: An Interview with John Carpenter, in John Carpenter (dir.), Prince of Darkness, Shout! Factory Collector's Edition Blu-ray (2013)

30. K. Thorne, *Black Holes and Time Warps* (W. W. Norton and Co., New York, 1994), pp. 483–491

31. T. Takács (dir.), The Black Hole, Millennium Films (2006)

32. A. Schapker, G. Godfree (script), Things You Can't Outrun, The Flash, season 1 (2014)

33. R.J. Sawyer, *Flashforward* (TOR, New York, 1999), p. 63

34. L. Zwerling, S. Hoffman (script), Countdown, FlashForward, season 1 (2010)

35. I.Ya. Aref'eva, I.V. Volovich, Time Machine at the LHC, ArXiv, https://arxiv.org/abs/0710.2696

36. R. Highfield, Time Travellers from the Future 'Could Be Here in Weeks', The Telegraph, https://www.telegraph.co.uk/news/science/large-hadron-collider/3324491/Time-travellers-from-the-future-could-be-here-in-weeks.html

37. D. Mandel, B. Kelley (script), Treehouse of Horror 23, The Simpsons, season 24 (2012)

38. E. Farhi, A.H. Guth, An Obstacle to Creating a Universe in the Laboratory. Phys. Lett. B **183**(2), 149 (1987)

39. E. Farhi, A.H. Guth, J. Guven, Is It Possible to Create a Universe in the Laboratory by Quantum Tunneling? Nucl. Phys. B **339**(2), 417–490 (1990)

40. A. Linde, Hard Art of the Universe Creation, ArXiv, https://arXiv.org/pdf/hep-th/9110037.pdf
41. G. Benford, *Cosm* (Orbit, London, 1998), p. 246
42. G. Benford, *Cosm* (Orbit, London, 1998), p. 368
43. S. Traweek, *Beamtimes and Lifetimes* (Harvard University Press, Cambridge, 1992), p. 33

8

Science Communication Redux: Returning to the Collision Point

8.1   Biting Ourselves in the Boson

In our previous discussion of scientific terminology salad in Chapter 4 we used an example from Gregory Benford's *Timescape* in which "cranks seemed to think constructing a new theory involved only the invention of new terms" [1]. The term *crank* is slang for someone with strange ideas or behaviors. A common synonym is *crackpot*, and both terms are casually tossed around by scientists when discussing individuals (usually nonscientists) who claim to have either disproven various laws of physics (most often related to the work of Einstein) or have original ideas that not only violate accepted science but appear to not have any consistent scientific or mathematical basis. As I noted in that chapter, such derogatory terms are counterproductive when debunking misconceptions, as they feed into the very backfire effects we must seek to avoid.

This issue is very important, so please bear with me as I further unpack it. As scientists, we need to be very careful when we glibly throw around such terms in public discourse. First of all, understand that I am not saying that we should simply ignore individuals who claim to have some new "scientific" discovery that subverts or supersedes the laws of nature as we know them. Quite the contrary; we should work to educate the public to be skeptical of all such claims and, taking a lead from Carl Sagan and the Committee for Skeptical Inquiry, debunk obvious pseudoscience wherever we see it. Our rationale is not to protect science from pseudoscience, but rather to protect the public from falling victim to pseudoscientific schemes that promise to

© Springer Nature Switzerland AG 2019
K. Larsen, *Particle Panic!*, Science and Fiction,
https://doi.org/10.1007/978-3-030-12206-5_8

stop the aging process, fix their arthritis with magnetic shoe insoles, or help them find a compatible mate through expensive personalized horoscopes.

But there is the wider issue of helping members of the public to hone their individual critical thinking skills so that when they are bombarded with statistics, competing interpretations, appeals to emotion, and what to them is scientific terminology salad, they will, at the very least, entertain the possibility that someone is trying to manipulate their opinions. More importantly, we want them to realize and accept that there are things that they simply don't understand, and that there is absolutely no shame in this, because no one knows everything (as our lack of a Theory of Everything proves). As college professors often tell their students, there is no such thing as a stupid question— if you feel the need to ask, it means you feel we have information to give to you. Put quite simply, we need the public to understand that they can come to scientists with questions, that we will not belittle them for asking these questions, and, finally, that they can trust the answers that we give to them. This may be a tall order, but in our increasingly technological and specialized society we can do nothing less. Such communication begins (and ends) with mutual respect.

Recall that debunking has the best chance of succeeding when it keeps its emphasis on a small number of key facts, avoids backfire effects (for example through ad hominem attacks), keeps explanations brief and focused (including why the misconception is wrong), and relies on visual aids (especially when dealing with numbers) [2]. For example, take the case of an individual making public claims of a supposed new scientific finding. As scientists, we all know that science is not a democracy; hypotheses are not voted upon by some Science Senate to turn them into a law of nature. Instead, papers are subjected to the peer review process, including submission to a journal for evaluation and, if they meet a certain bar of scientific rigor, publication. However, the process does not end there; after publication, other scientists will continue to evaluate the claims, and continue to ask "what if", in some cases extending the ideas to additional situations and examples or pointing out shortcomings in the original analysis. It's a messy, long-winded, never-ending process, but when it works (thankfully most of the time), it improves our understanding of the universe. Every time a scientist shares an idea or experimental result with the larger scientific community it becomes a target for others to try to poke holes in it. Sometimes there are relatively simple errors or oversights made that can be fixed, and a revision issued, for example the statistical interpretation problems found by Adrian Kent in two of the RHIC related safety reports. In severe cases papers are even retracted, either because a large enough error is discovered after publication or, most embarrassingly for the scientific establishment, it is discovered that results were faked.

This is the self-correcting way the system worked, at least before the Internet. The availability of free preprints (especially on arXiv) allows authors to receive valuable feedback before seeking publication in peer-reviewed journals, saving them possible embarrassment and assuring that ideas are posted in a timely manner (establishing primacy in the case where several individuals are working on a topic at the same time). Given the high price of subscriptions to peer-reviewed journals, preprint services such as arXiv also provide access to new theoretical ideas and experimental results to those with limited financial resources (as well as the general public). However, there have been criticisms of arXiv's apparently unevenly applied rejection policies [3] and an acknowledgement from the scientific community that there are preprints on arXiv that would have been weeded out by a normal peer review process. Journalists and the public read these archives, but are they well situated to tell the difference between an original and groundbreaking result, a highly speculative idea that is nonetheless grounded in accepted science, and sloppy research or erroneous interpretations that can be easily debunked by experts in the field? I think we can all agree that the answer is no. The issue is even more complicated due to the existence of other online preprint archives, such as viXra.org (arXiv spelled backwards), which proudly promises to post *any* paper submitted [4]. This is the reality of the scientific supermarket of ideas in the 21st century.

So what are individual scientists and accelerator press officers to do to combat the blatant misconceptions and speculative preprints that don't conform to scientific standards of analysis and evidence? If the goal is to deescalate public fears and promulgate an environment of trust, using derogatory terms to describe LHC critics is, shall we say, less than helpful. For instance, take CERN scientist Lyn Evans' characterization of Otto Rossler as "a 'crazy' retired professor". Evans also offers that "We have shown him where his elementary errors are, but of course people like that just will not listen" [5]. While Evans' palpable exasperation is understandable, the readers of *The Daily Mail* might be put off by his curt and seemingly close-minded dismissal of the chemist. Again, I am in no way defending Rossler's arguments, but rather noting that it is possible to criticize the ideas (the science, or lack thereof) while not attacking the person. CERN physicist John Ellis espouses a similar view in his CERN in-house colloquium on LHC safety. During the Q&A an audience member voices frustration at the public's fear and blames it on a lack of understanding of science, sensationalizing by the media, and journalists editing scientists' comments and taking them out of context. He also asserts that CERN

Fig. 8.1   John Ellis (User Alban, CC-BY-SA-3.0, via Wikimedia Commons)

scientists need to criticize these "crazy people" (meaning people who bring up safety concerns), including those who are fellow scientists (including Rossler). Ellis' measured response should be a lesson for us all. He clearly notes that he had not directly criticized Rossler's work and flatly states that "no one at CERN should make such criticisms" [6] (Fig. 8.1).

On the other hand we have a pithy quote given to *The Telegraph* by "rock-star" physicist Brian Cox in early September 2008: "*Anyone who thinks the LHC will destroy the world is a twat*" [7]. While astrophysicist and popular science writer Ian O'Neill had initially praised Cox's comment on his own blog, declaring it "Superb", two days later O'Neill offered a clarification after Cox's own walk-back. In response to being called "rather abusive to people who are worried about the LHC", Cox posted a discussion forum statement clarifying that he doesn't "think that people who are worried about new scientific endeavors are 'tw*ts'", calling skepticism a "much-needed skill" [8]. He then reiterates very briefly that the energies of the LHC are lower than nature achieves (meaning cosmic rays) and are therefore harmless. Cox then follows some of the suggested best practices for debunking by offering a limited number of clearly articulated points with the intention of replacing misconceptions with logical alternatives. First, he suggests that the public should trust particle physics experts on the safety of the LHC just as we trust the experts who design airplane wings. He also makes the point that particle physicists are just as concerned for their own families' safety as are members of the general public. Cox then refers to Otto Rossler as a "distinguished biochemist" who nevertheless "based his argument on a pretty basic error in General Relativity" [8]. Compare this clear, straightforward, respectful refutation of Rossler's work (not the man himself) to Evans' characterization. Cox carefully notes that he is not stating that Rossler should not share his ideas on the subject, but rather that "wisdom comes from noticing when ones [sic] opinion is disproved by evidence. This is the key to science" [8].

O'Neill was far less conciliatory on his blog. While admitting that Cox's original statement "could be misconstrued as being offensive", he thought "the vast majority understood what he was saying" [8]. This begs the question, the majority of whom—scientists, members of the general public, or those with concerns about the LHC? Again, the frustrations voiced by Cox and other physicists over both blatant misconceptions and fear-mongering promulgated on the Internet (as well as the resulting threats made against physicists themselves) are certainly justified. But the central lesson of this very teachable moment is to see how emotional responses are not helpful in swaying public opinions about science and scientists. In light of this discussion and the knowledge gained in our survey of the topic over the previous seven chapters, we now turn to further examples of statements issued by CERN and its member scientists for the purpose of calming fears and engendering goodwill.

8.2   CERN Communications: The Good, the Bad, and the Funny

There is perhaps no more obvious place to begin than the official CERN page entitled "The Safety of the LHC" [9]. Both the media and general public are important consumers of the content of this information (particularly as it comes up among the top sites in a Google search of "lhc safety" or "is the LHC safe"). However, "safety" is not a topic listed under the "Resources" tab on the CERN website, nor does this particular page come up when one selects "General public" as an audience under the same tab.

Without dwelling too much on web design, the initial sense of the page is quite simply massive overload. Running a full six screen lengths of pure text with no explanatory diagrams, the page is markedly different in style and tone compared to other CERN webpages. The LHC Safety page begins with a reaffirmation by the LHC Safety Assessment Group (LSAG) of the safety of the LHC and its experiments, noting that nature has already done whatever the LHC can accomplish (a reference to high-energy cosmic rays). A brief refutation of specific danger scenarios (including microscopic black holes, strangelets, and vacuum bubbles) follows, largely summarized from the 2008 John Ellis, Gian Giudice, Michelangelo Mangano, Igor Tkachev and Urs Wiedemann LSAG Safety Assessment Group report. A link to the complete report is also included. Those interested in details are encouraged to read the report itself as well as its references. The summaries are clearly written for the nonscientist but contain very few explanatory links to additional information.

The remainder of the page is material that is best described as addenda meant to bolster the case for the LHC's safety, but the presentation directly contradicts the admonitions by Cook and Lewandowsky to avoid offering too many key points that can overwhelm the intended audience. A member of the general public who is leaning towards cynicism might misconstrue this as a case of CERN protesting too much. This additional material could have easily been briefly summarized with links provided to further information for interested readers.

Special emphasis is given in the ancillary material to the peer-reviewed publications of the LSAG report and the paper "Astrophysical Implications of Hypothetical Stable TeV-scale Black Holes" by Steven Giddings and Michelangelo Mangano. It is correctly explained that these papers have undergone rigorous review by scientists from around the world and have been endorsed by experts in the related fields "including several Nobel Laureates in Physics. They all agree that the LHC is safe" [9]. Examples of those endorsements run parallel to this explanation in another column of text (interestingly done in a fainter shade of gray). Again, a skeptic might consider it a conflict of interest when one of those quoted Nobel Laureates is the same Frank Wilczek from the *Scientific American* editorial incident. Confusion might be created by the fact that another highlighted quote is from Steven Hawking), who, despite the quotation listed here that "The LHC is absolutely safe", has made comments elsewhere that have raised fears in the press (see Chapter 6). The first laureate quotation, by Vitaly Ginzburg, also descends into ad hominem attacks, referring to those who claim that the LHC could produce "dangerous black holes" as "unqualified people seeking sensation [sic] or publicity" [9].

Additional works supporting the argument that there is no concern about black holes are then cited before criticisms of Otto Rossler and Rainer Plaga are addressed. In the case of Rossler these criticisms are presented with perhaps more vehemence (and space) than is helpful; recall that Cook and Lewandowsky recommend avoiding the familiarity backfire effect, in which discussing the misconception (here the claims not supported by accepted science) too often lead to it being more well-known by the public than are the established facts. The page ends with an appended list of links including a "further expert comment" and "another independent assessment" [9]. Again, while this information could be useful to the interested reader, in its current arrangement (especially at the end of six screens of text), it could be overlooked by some who would benefit the most from it. The lesson is that sometimes less is more, especially when it is clearly focused in a manner that speaks to its intended audience.

Having criticized one of the CERN communication machine's products, it is only correct that I laud them for others, because CERN has certainly gone above and beyond in its attempts to make information about the LHC available to the public in an engaging and level-appropriate manner, for example the *Angels & Demons—The science behind the story* webpages described in Chapter 6. Another specific example is their response to a conspiracy claiming that the LHC would open up a portal to hell in September 2015, heralding the end of the world. NASA became involved in debunking the same end of the world prediction, in particular claims that a killer asteroid would impact in the Caribbean during the last two weeks of that month. In response, NASA issued a statement on August 19, 2015 with the curt title "There is no asteroid threatening earth" [10]. As straightforward as this might seem, recall the familiarity effect. The hasty reader of the story and its opening paragraph could easily over look words like "no" or "erroneously claiming" and instead fixate on the threatening asteroid part (Fig. 8.2).

Noting that this "in-your-face" response by NASA seemed to produce the unintended effect of increasing the online notoriety of the claim, CERN spokesperson Arnaud Marsollier and CERN social media specialist Kate Kahle instead sought a different approach, one more aligned with Cook and Lewandowsky's recommendations [11]. Instead of issuing a press release or posting a specific statement on their website, the question "Will CERN open a door to another dimension?" was listed in the middle of a collection of other conspiracy related questions (such as "Does the LHC trigger earthquakes?"). The straightforward answer "CERN will not open a door to another dimension" is followed by a brief explanation that experiments at the LHC might instead provide evidence of the existence of extra dimensions. A link to more

Fig. 8.2 Artistic representation of an apocalyptic asteroid impact (Don Davis/NASA, public domain, via Wikimedia Commons)

information is also included. Note that the words "portal to hell" or "end of the world" are not used, and that the brief, pointed answer directs the reader's attention toward the facts rather than reinforcing the misconception. The tactic was deemed successful, as the page was at the top of the results in Google searches for "Sept. 23 CERN" [12].

Another approach used by CERN and its employees (both officially and unofficially) has been humor in the form of parodies of CERN press releases and song parodies. Most notable among the latter is "The LHC Rap", the brainchild of then CERN Communications intern Katie McAlpine. Released on YouTube in July 2008, the video, which features Katie and fellow interns lip-synching and dancing among the LHC detectors, now has over 8 million hits. The catchy ditty explains in brief how the collider works and what each of the major detectors at CERN hopes to observe, including dark matter, the Higgs boson, and evidence of extra dimensions.[1] Another particularly brilliant example is a 2015 parody of Howie Day's "Collide", written by U.S. Communications Manager for the LHC experiments Sarah Charley and three graduate students. Written from the POV of a proton in the LHC,[2] the music video so impressed Day that he asked to have a tour of the facility. In return, he made a special music video for the song during his visit.[3] It is truly disappointing that Day's video has fewer than 50,000 hits on YouTube and the original only has a third of that, as the parody is a highly creative and engaging description of the LHC experiments. Both this project and "The LHC Rap" also paint a picture of particle physicists that significantly deviates from the classic nerd stereotype described in Chapter 2. While a few of the senior scientists have brief cameos, the vast majority of the CERN students and employees in the videos are hip, young, multiracial, men and women who are depicted as not only intelligent but creative and enthusiastic. These positive public ambassadors for CERN and for particle physics in general should be encouraged to continue to channel their talents in this type of public outreach. This is especially important because creative works that target a younger audience (including the next generation of potential scientists currently in high school or even earlier) could go a long way to preventing unnecessary misconceptions and fears among the next generation.

Another parody that potentially targets a younger generation is the zombie film *Decay*. The creation of then University of Manchester Ph.D. student Luke Thompson and his friends and colleagues, *Decay* was brought to life over several years and for around $3000. Thompson was motivated to take a satirical look at the LHC end of the world fears by incorporating "intentionally bad science" as well as other tropes such as mad scientists. Thompson proudly notes that his fictional "scientists are even worse than the bad scientists in Hollywood movies" [13]. However, some caution is required here. Tibi Puiu notes that the film is true to some of the science at CERN despite the over-the-top aspects of the "Higgs radiation" and, of course, the zombies, and is therefore an "effective medium for popularizing" the LHC [14]. But can the average viewer recognize the difference between the "intentionally bad" and the accurate science? Of course, this speaks to issues of science literacy once again; we would certainly hope that the audience could tell the difference, but

---

[1] https://www.youtube.com/watch?v=j50ZssEojtM
[2] https://www.youtube.com/watch?v=-1AF7GwAxfI
[3] https://www.youtube.com/watch?v=1YB0xM9cgr8

Fig. 8.3 Zombies overrun CERN in *Decay* (2012) (H2ZZ Productions, CC-BY-NC; screen capture by author) (Public domain)

what are the consequences of failing to do so? In the case of *Decay*, it might be to believe that the Higgs particle is somehow dangerous, that CERN is involved in conspiracies, or that the valid safety concerns of scientists would be overruled by management.

Thompson's colleague Clara Nellist, also involved with the film, voiced hopes that the film would promote conversations about physics. She notes that her grandmother had even asked her about the physics in the film [13]. What about those viewers who do not have a physicist family member? While the introductory disclaimer calling the film "purely a work of fiction" should suffice to dissuade misconceptions, those who are susceptible to believing in conspiracies might take it completely differently. I would therefore suggest that there is a possibility that the use of stereotypes of not only mad scientists and conspiracies, but also the Higgs boson being dangerous, feeds into rather than ameliorates distrust of the LHC. The parody is subtle here, not blatantly humorous (as in *Spaceballs* or *Galaxy Quest*) and may be lost on some viewers. The use of stereotypes and tropes also feeds into the familiarity backfire effect. It should be noted that, although CERN did not authorize or endorse the film, they were aware of the project. The film is available online for free under a Creative Commons license and selected scenes could easily be used in college classes (and elsewhere) to facilitate discussions of many of the issues discussed in this book (Fig. 8.3).[4]

---

[4] https://www.decayfilm.com/

Another example of CERN humor is a series of April Fool's Day parodies of news releases that sometimes walk a fine line between obvious satire and more subtle ribbing. Posted on the official website and announced on Twitter and other social media, a disclaimer is posted the next day along with an invitation to explore real science at CERN (with a link). For example, when April 1, 2018 fell on Easter a fake news release announced the discovery of the "Eggeron" or "Humpty Dumpty particle" (complete with gratuitous egg related puns, such as referring to "hard-boiled scientists") [15]. My personal favorite of these parodies is the clever April 1, 2015 fake press release claiming that CERN had discovered evidence of the Force. *Star Wars* references abound, including an unnamed "diminutive green spokesperson" who speaks suspiciously like Yoda. A "rogue researcher" named Dave Vader, drawn to the "Dark Side of the Standard Model", is said to be considering building his own research station [16]. There are also photoshopped images of scientists using the Force. This playful geekiness is endearing, but is playing to a particular audience (*Star Wars* fans) who are potentially already more science-savvy than the general public.

On the other hand, the 2017 April Fool's press release is less clearly playful and can be viewed as directly poking fun at conspiracies claiming that the so-called Face on Mars feature is an artificial alien structure. In this spoof it is claimed that Olympus Mons, the largest volcano on Mars, is the remains of a millions of years old particle accelerator. References to the movie *Stargate* are sprinkled throughout the article, especially in the names of the fake scientists (e.g. Fadela Emmerich, named for the director Roland Emmerich; Friedrich Spader, a reference to actor James Spader, who portrayed Egyptologist Daniel Jackson; and Eilert O'Neil, a reference to Kurt Russell's character Jack O'Neil) [17]. Not only could some individuals be put-off by the parody (interpreting it as poking fun at a belief they hold), but the story could actually cause individuals to question its possible truth. In the most extreme case, those who hold such conspiracy beliefs could (and have) used it as evidence, especially because in this case the usual April Fool's disclaimer came on April 4 rather than April 2 (Fig. 8.4).[5]

The most effective type of humor is perhaps that turned inward, self-effacing rather than poking fun at others. A prime example is CERN's 2014 April Fool's parody of the social media outcry of its use of the much-maligned Comic Sans font in the PowerPoint slides for the historic July 4, 2012 announcement of the discovery of the Higgs particle [18]. In this faux

---

[5] https://www.youtube.com/watch?v=3qUluLWj4lY;   https://www.quora.com/Is-it-true-that-CERN-found-an-ancient-particle-accelerator-on-Mars

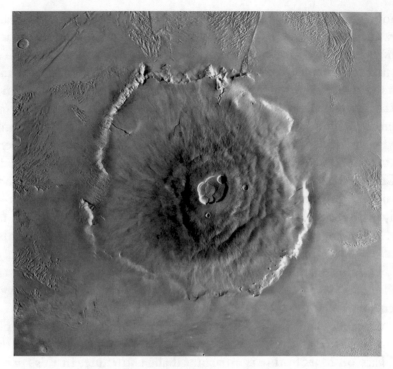

Fig. 8.4 Martian volcano Olympus Mons (Image by NASA, modifications by Seddon, public domain, via Wikimedia Commons)

announcement it is claimed that all CERN communications would be switching to the hated font in order to update the laboratory's image. The fictional "presentation analyst May Dupp, who worked closely with comic-book artists and circus clowns" is quoted as noting that the success of the Higgs particle discovery announcement was mainly due to the font and not the physics content. Other supposedly adopted changes to CERN's image were "adding a selfie of Justin Bieber to the CERN logo and rebuilding the LHC in the shape of a triangle", as well as adding "animations of little clappy hands" to important announcements posted online. While the mention of the CERN logo may be a reference to conspiracies surrounding the logo (see Chapter 4), it is not an obvious slight towards such believers (as suggesting the logo be converted to a series of 7's, for example, would have been) [19].

Finally, although not a true parody, Ph.D. student Nathan Readioff's Lego LHC (miniature models of four of the detectors connected by dipole magnets) deserves special mention. Clearly a labor of love, Readioff's petition (with CERN's approval and help) to become an official Lego product was unfortunately not approved, despite the clever accompanying YouTube video in the form of a parody of an iPad commercial.[6] Despite this disappointment, the design (utilizing a mere 317 pieces) is available online (miniature Higgs boson not included).[7] From zombies and Legos to raps and parodies, CERN scientists, students, and staff have proven that they know how to bring particle physics to the general public in an engaging manner; they just don't always remember to invoke best practices. While they creatively make the case that the LHC is both exciting and safe, they ignore one important fact—it is also very, very expensive. CERN's annual budget is about a billion U.S. dollars per year, and the total cost for finding the Higgs boson is estimated to have been about thirteen billion U.S. dollars [20]. Part of the public outreach responsibility of scientists needs to be a convincing explanation of *why* countries and consortiums should dedicate so many resources to discovering microscopic particles.

8.3   Making the Case: Why Do We Need the LHC?

In the 2011 episode "The Mermaid" of *The Boat*, Bubble asks Julia what particle colliders are used for. She explains that they are "used to make very little things, called hadrons, crash into each other" in order to "find what people call the God Particle. A particle which can be used to create matter. Anything.... Things that we don't know. Everything from nothing". Bubble asks "Like a magician?" to which Julia, obviously pained, answers "Like God" [21]. Recall that this idea of playing God is central to the Frankenstein stereotype and the mad scientist trope that are so deeply ingrained in our society's view of the worst of science. Not only do particle physicists probe the early universe (creation itself), but they freely celebrated their ability to create the "God Particle" in the LHC. Is it any wonder that people question why we need larger and larger particle accelerators? Is it a matter of megalomania (to become more Godlike) or the Freudian interpretation of wanting an instrument of power that is the biggest possible?

   In the particle physics based fiction we have examined throughout this book, the purely scientific search for knowledge for knowledge's sake is seldom the main impetus behind fictional accelerator experiments. Instead, the goal is more commonly national pride (or the personal glory of a Nobel Prize) or a search for limitless clean energy. But this last reason can also be either an

---

[6] https://www.youtube.com/watch?v=iuG6ZJFC6qk
[7] https://ideas.lego.com/projects/5c3aec53-00d2-40a2-be73-9e2db09da86f

admitted ruse for public consumption (as in the case of *The Void*) or subverted by the influence of others into a weapon (e.g. *Angels & Demons*). While antimatter cannot be used as either an energy source or a weapon because too little is made and it requires far too much energy to produce in the first place, what of the general physical principles related to particle accelerators? Is the idea of the boson bomb in *A Hole in Texas* so ridiculous given the power released when a neutron collides with a uranium or plutonium nucleus in the heart of an atom bomb? Robert Wilson and Raphael Littauer explain that experiments up to an energy of about 100 MeV deal with the properties of the nucleus as a whole, including the release of energy in fission. But at higher energies it is the properties of the individual particles (such as colliding protons and the quarks that make them up) that are the focus of the collisions. Wilson and Littauer note that there is no evidence that great amounts of energy await us within individual protons or neutrons [22]. Again, it requires a tremendous amount of energy to do these experiments, making them lousy energy sources or weapons. Industrial uses of smaller accelerators includes producing radiation for the purpose of making materials such as plastics or rubber harder, sterilizing food, or nondestructive testing of critical metal components. X-ray machines in hospitals and airports as well as studies of the crystalline structure of proteins and other large molecules also utilize or benefit from principles of accelerator physics [23].

However, what of particle physics' obsession with bigger and higher energy particle accelerators that have clearly exploratory rather than industrial or medical uses? How does one justify the expenses associated with these mammoth projects (of course, beyond the economics of the jobs they create)? The "CERN answers queries from social media" webpage offers that "CERN exists to understand the mystery of nature for the benefit of humankind" [24]. Putting aside those individuals who would suggest that the "mystery of nature" is either a theological question, or beyond the purview of human minds to contemplate, it is fair to ask just what is this "benefit" that CERN is offering humanity? As Brian Cox notes, "we scientists will always be called upon to justify our voyages of discovery in economic language simple enough (dare I say it?) for politicians to understand" [25].

In 1985, Sheldon Glashow and Leon Lederman tried to do just that, specifically as applied to the then planned SSC. Their argument boils down to four points: "challenge, spinoff, pride, and duty" [26]. Note that the first two are more concrete reasons, while the remainder are philosophical. Meeting the technological challenges posed by building high-energy accelerators such as SSC and LHC requires innovation and creativity that have applications to manufacturing, planning, and communications processes. The most famous

Fig. 8.5 The first server on the World Wide Web (Coolcaesar, CC-BY-SA-3.0, via Wikimedia Commons)

example related to CERN is undoubtedly the World Wide Web. Spinoff technologies are an often-cited tangible benefit of such large-scale projects (for example, in the case of the space program). Advances in computing, superconductors, and detector technology have resulted from building the LHC, with applications to medicine, security, and other fields [27]. Another concrete benefit from CERN and other particle accelerator facilities is education. For example not only do CERN experiments provide data to Ph.D. students from around the world, but CERN offers summer programs for teachers and students as well as internships (Fig. 8.5).

There is still the issue of the more esoteric benefits, pride and duty. National pride was clearly a motivating factor behind the building of the SSC (and selling the idea to the government and public), to return particle physics primacy to the United States from Europe much in the same way that John F. Kennedy challenged the U.S. to win the Space Race by being the first to put humans on the moon. Glashow and Lederman offered in 1985 "When we were children, America did most things best. So it should again" [28]. In invoking "duty" as a reason to build large accelerators, Glashow and Lederman refer to our role as members of the universe as a whole, opining that "it is our sacred duty to know its deepest secrets, as well as we are able". In the end it is "simply *the need to know* that compels us to build a bigger and better accelerator" (emphasis original) [28]. This last reason is the one that truly drives particle physicists. It is basic human curiosity, the insistent query of a young child—"why"?

In his commentary "In Defense of the LHC", Brian Cox refers to a 1966 speech by physicist Hendrik Casimir to the U.S. Commerce Department in which Casimir reflects on the fact that it was so-called pure research (based on the need to understand reality rather than the desire to build a better mouse trap) that led to many of the important technologies of his century, including transistors and radio communications [29]. Dr. Asa Breed of Kurt Vonnegut's *Cat's Cradle* echoes this sentiment, speaking for many scientists who feel pressure to justify what they do in terms of concrete returns on financial investments: "I'm sick of people misunderstanding what a scientist is, what a scientist does.... In this country most people don't even understand what pure research is.... It isn't looking for a better cigarette filter or a softer face tissue or a longer-lasting house paint" [29].

As physicist Robert Wilson offered to Senator John Pastore during a 1969 Congressional hearing, using particle accelerators to explore the nature of reality "has nothing to do directly with defending our country except to make it worth defending" [30]. In an age of stagnant science literacy and increasing national debt, where blatant misinformation and claims of nefarious conspiracies spread across the Internet without respect for national borders, it is more important than ever that scientists take responsibility for being their own best advocates, presenting a clear and convincing case for the precious resources they need to continue exploring the wonders of the world around us. They must learn to effectively communicate with nonscientists whose tax dollars (or pounds, or euros) fund those experiments. In response to the need for increased effective communication between scientists and the general public despite the fact that most scientists have no training in this area, many professional groups, such as the AAAS, American Chemical Society, and Union of Concerned Scientists, offer resources and training, both online and in the form of workshops and internships. Those who have a special interest in and talent for communication can even become science communications specialists, working at CERN and other science facilities. These professionals have a difficult but vitally important job, bridging worlds and speaking two very different vernaculars, and as such are indispensable members of the wider research team. Not only do they write press releases and social media posts, answer questions from the press and general public, they also plan exciting outreach events that bring science to the widest audience possible. From public lectures and tours to local exhibitions and even global celebrations (for example International Dark Matter Day[8]), public outreach is central to their sphere of influence.

However, individual physicists cannot abandon their personal responsibility to play a role in this endeavor, especially those employed at smaller institutions that do not have dedicated professional science communication specialists. We all have a responsibility to share our personal scientific autobiographies with the public, in whatever ways our talents allow, and always embracing best practices. It falls upon us to share with the world our passion for the grand adventure that is the search for the Theory of Everything.

References

1. G. Benford, *Timescape* (Bantam Books, New York, 1980), p. 235
2. J. Cook, S. Lewandowsky, *The Debunking Handbook, version 2* (University of Queensland, St. Lucia, 2012), p. 3
3. Z. Merali, ArXiv Rejections Lead to Spat Over Screening Process, Nature, https://www.nature.com/news/arXiv-rejections-lead-to-spat-over-screening-process-1.19267
4. G. Brumfiel, What's ArVix Spelled Backwards? A New Place to Publish, Nature, http://blogs.nature.com/news/2009/07/whats_arXiv_spelled_backwards.html
5. J. Petre, Meet Evans the Atom, Who Will End the World on Wednesday, The Daily Mail, http://www.dailymail.co.uk/sciencetech/article-1053091/Meet-Evans-Atom-end-world-Wednesday.html
6. J. Ellis, CERN Colloquium: The LHC Is Safe, CERN, http://cds.cern.ch/record/1120625?ln=en
7. I. O'Neill, Anyone Who Thinks the LHC Will Destroy the World Is a Twat, Astroengine, https://astroengine.com/2008/09/04/brian-cox-anyone-who-thinks-the-lhc-will-destroy-the-world-is-a-twat/
8. I. O'Neill, A Statement by Professor Brian Cox, Astroengine, https://astroengine.com/2008/09/06/a-statement-by-professor-brian-cox/
9. The Safety of the LHC, CERN, https://home.cern/science/accelerators/large-hadron-collider/safety-lhc
10. NASA: There Is No Asteroid Threatening Earth, NASA, https://www.nasa.gov/jpl/nasa-there-is-no-asteroid-threatening-earth
11. J. Letzing, CERN Is Seeking Secrets of the Universe, or Maybe Opening the Portals of Hell, Wall Street Journal, https://www.wsj.com/articles/cern-is-seeking-secrets-of-the-universe-or-maybe-opening-the-portals-of-hell-1459800113

---

[8] https://www.darkmatterday.com/

12. R. Mandelbaum, Debunking Doomsday at CERN, Again, Scienceline, http://scienceline.org/2016/01/debunking-doomsday-at-cern-again/
13. M. Reisz, Students Fight for Their Lives as Living and Dead Collide, Times Higher Education, https://www.timeshighereducation.com/students-fight-for-their-lives-as-living-and-dead-collide/422289.article
14. T. Puiu, CERN Scientists Direct and Release Zombie Movie, ZME Science, https://www.zmescience.com/other/cern-zombie-movie-decay-043243/
15. F. Mazzotta, K. Kahle, 'Humpty Dumpty' Particle Discovered, CERN, https://home.cern/about/updates/2018/04/humpty-dumpty-particle-discovered
16. C. O'Luanaigh, CERN Researchers Confirm Existence of the Force, CERN, https://home.cern/about/updates/2015/04/cern-researchers-confirm-existence-force
17. A. Marsollier, Ancient Particle Accelerator Discovered on Mars, https://home.cern/about/updates/2017/04/ancient-particle-accelerator-discovered-mars
18. R. Urquhart, CERN Higgs Boson V Comic Sans Debacle, Huffington Post, https://www.huffingtonpost.co.uk/robert-urquhart/cern-higgs-boson-comic-sans-_b_1148058.html
19. C. O'Luanaigh, CERN to Switch to Comic Sans, https://home.cern/about/updates/2014/04/cern-switch-comic-sans
20. A. Knapp, How Much Does It Cost to Find a Higgs Boson?, Forbes, https://www.forbes.com/sites/alexknapp/2012/07/05/how-much-does-it-cost-to-find-a-higgs-boson/
21. M. Cistaré, J. Naya (script), The Mermaid, The Boat, season 2, (2011)
22. R.R. Wilson, R. Littauer, *Accelerators: Machines of Nuclear Physics* (Anchor Books, Garden City, 1960), pp. 167–168
23. A. Sessler, E. Wilson, *Engines of Discovery: A Century of Particle Accelerators,* Rev. edn. (World Scientific, Singapore, 2014), p. xiv–xv
24. CERN Answers Queries from Social Media, CERN, https://home.cern/resources/faqs/cern-answers-queries-social-media
25. B. Cox, In Defense of the LHC, Popular Science, https://www.popsci.com/sci-tech/article/2008-09/defense-lhc
26. S.L. Glashow, L.M. Lederman, The SSC: A Machine for the Nineties. Phys. Today **38**(3), 32 (1985)
27. S. Carroll, *The Particle at the End of the Universe* (Plume, New York, 2012), pp. 274–275
28. S.L. Glashow, L.M. Lederman, The SSC: A Machine for the Nineties. Phys. Today **38**(3), 34 (1985)
29. K. Vonnegut, *Cat's Cradle* (Dial Press, New York, 2010), pp. 40–41
30. E. Kolbert, Crash Course, The New Yorker, https://www.newyorker.com/magazine/2007/05/14/crash-course

9

Conclusion: Breaking the Chain (Reaction)

We have discussed myriad topics in this work, from bosons and black holes to zombies and interdimensional monsters. We now return to the points raised in the introduction, hopefully with our eyes opened to the idea that the reason why some people find science frightening is precisely the same reason why others find it the thrill of a lifetime—the unknown. How we communicate about our investigations of this unknown has the potential to send someone from one side of the fear/awe barrier to the other, because, as we have seen, words matter.

Unfortunately, scientists aren't always as precise with their language as they are with their mathematics. As Lloyd Simcoe notes in the novel *Flashforward*, CERN scientists have "been saying outrageous things… deliberately using loaded words, making the public think we're doing things that we aren't" [1]. Compare this fictional dialogue to the following admonition by the very real CERN physicist John Ellis that particle physicists should "recalibrate our rhetoric. We talk about recreating the Big Bang, which in some narrow technical sense is correct. But a lot of people are going to misunderstand this and think we're really going to be" [2]. In the Age of the Internet, hyperbole leads to misconceptions, which fuels conspiracy theories, leading to mistrust, and calls for restrictions on the scientific endeavor. Scientists simply need to stop giving their critics ammunition.

This push for physicists to effectively communicate with the general public is certainly not a new idea. For example, in 1996, long before the first RHIC safety study and while the LHC was still in the planning stages, nuclear physicist Ruth Howe warned that "We physicists can no longer afford the luxury of talking mainly to ourselves…. Physicists must involve the media and the

© Springer Nature Switzerland AG 2019
K. Larsen, *Particle Panic!*, Science and Fiction,
https://doi.org/10.1007/978-3-030-12206-5_9

public they serve with physics and its exciting results" [3]. While this call to action was undoubtedly answered by some individuals, a 2003 poll of scientists found that 42% did not engage in any public outreach. When asked why they did not participate in public outreach (giving them the choice of selecting multiple answers) 76% cited lack of time, while 28% lacked the desire, and 17% simply "did not care" [4]. That scientists felt that they lacked the time to engage in productive public outreach should come as no surprise, given the pressure to secure funding (which can be an extremely time-intensive process) and satisfy other job requirements. In the case of academics, there has historically been a mantra of "publish or perish", meaning that the reward system of promotion and tenure values peer-reviewed publications (especially in so-called high impact journals) over quality teaching or service. Even within the already depreciated service category, university-wide committee work is often more highly valued than public outreach [5]. Given this reality, it is no wonder that the suggestion of a scientist working with the media rather than serving on NSF grant review committees, giving public talks rather than technical colloquia, or writing popularizations of one's work rather than a technical preprint (with the hopes of turning that into the next peer-reviewed paper) would be met with equal parts horror and incredulity.

While there are only so many hours in a day, and scientists do not have superpowers outside of science fiction, perhaps integrating these tasks *with* our scientific work (rather than in addition to or instead of) is in order. Communicating with a general audience not only hones our writing and speaking skills, but forces us to consider aspects of our research that we might not otherwise ponder, such as how our work fits historically with similar research, or why someone outside of our field should be interested in what we have done. In communicating with those outside of our highly specialized silo of knowledge (not just the general public, but professionals in other disciplines) we may find important interdisciplinary connections, and perhaps even discover an extension of our research, or a new avenue for consideration. Our teaching may directly benefit from these endeavors, and, if none of these reasons is sufficiently motivating, consider that scientists and their graduate students would all benefit from more clearly written peer-reviewed papers. Harvard trained neurobiologist turned science communicator Mónica I. Feliú-Mójer suggests that helping scientists to communicate more effectively with the public will benefit the scientific establishment at large, leading to more effective grant applications, journal articles, and, perhaps most importantly, it will improve science education, leading to higher rates of attracting and retaining students in STEM fields. In her words, it "can help make science more diverse and inclusive" [6].

The devaluing of outreach, in particular written popularizations, is an important issue, and a long-standing snobbery within professional science. While some well-known scientists took an active role in the popularization of science in the early 20th century, for example British astronomer Sir James Jeans, the prejudices we see today were already in place at that time. In his *Discourses: Biological and Geological Essays* (1898), Thomas H. Huxley warned that success in the popularization of science "has its perils for those who succeed. The 'people who fail' take their revenge… by ignoring all the rest of a man's work and glibly labeling him a mere popularizer" [7]. In the late 20th century, those scientists who chose to speak with the general public found themselves *Saganized*, a pejorative term referring to Pulitzer Prize winning astronomer Carl Sagan, who suffered backlash from the scientific community for his media efforts to bring science to the average person. In response to such prejudices, there is an effort to replace the term *popularization* with *expository science*, a term with less political baggage [8] (Fig. 9.1).

Fig. 9.1   Carl Sagan (NASA/JPL, public domain, via Wikimedia Commons)

Regardless of how we refer to the process, the prejudice itself will not be overcome so long as we embrace a misconception held by too many in the scientific community, namely that popularization equals the simple watering down of pure science in order to make it palatable to the general public [9]. As anyone who has written such works can attest, it is not that simple. Instead, the effective popularization of science requires two skill sets rather than one; not only does the author have to possess an understanding of the science at hand, but the effective communication skills to *translate* the science for a nontechnical audience. As a result, there is not only room for, but a *need* for, both those who create the scientific knowledge and those who more widely disseminate it beyond the technical discipline. This is the reason for the need for highly qualified science communication specialists. As astronomer Margaret Lindsay Huggins wrote of such work in her own field in 1907,

> The mission of these special workers is to collect, collate, and digest the mass of observations and papers; to chronicle, in short, on one hand, and, on the other, to discuss and suggest, and to expound; that is, to prepare material for experts, to inform and interpret [for] the general public. There is urgent need of a better-educated public opinion in this country. [10]

These words still ring true today.

There are also myriad other forms of science communication besides strictly nonfiction popularizations. For example, scientists should be encouraged to write more novels and screenplays, hopefully with a mind to avoid unproductive stereotypes of not only scientists, but the critics of science (and the relationship between science and religion). As Kip Thorne notes of his role in the development of the screenplay (and final cinematic version) of *Interstellar*, "most important to me was our vision for a blockbuster movie *grounded from the outset in real science*" (emphasis original) [11]. Authors of science fiction (including both scientist-authors and authors who simply love science) provide a valuable service when they seize the teachable moment and in an appendix or reading list offer their readers further means to explore the science that influenced the novel they have just enjoyed reading. Franklin Clermont's *The Voices at CERN* ends with a reading list of popular level works, including Sean Carroll's *The Particle at the End of the Universe*, Amir D. Aczel's *Present at Creation*, Ian Sample's *Massive*, and Leon Lederman's *The God Particle*, all of which have been cited in this book. In his afterword Clermont explains how he tried to weave together real science and speculative, philosophical questions, and hopes that the reader not only learned some science while reading his work, but will continue to monitor the media for new developments from CERN. Likewise Richard Cox explains in his author's note to *The God Particle* that his novel was inspired by Leon Lederman's *The God Particle*. He also gives a shout out to John Cramer's novel *Einstein's Bridge* for its description of what the SSC might have been like. In his afterword to *Cosm,* scientist-author Gregory Benford briefly discusses the science behind his plot device of the child universe and points readers to Alan Guth's writings on the subject (both the more and less technical versions).

On the other hand, when a writer attempts to ground their fictional work in real science, it would be helpful if they could be clearer (more honest) in not presenting science fiction as science fact. For example, the Coda to Christie Golden's *Shadow of Heaven* is presented as the final statement of renegade Romulan scientist Dr. Telek R'Mor. The fictional science of this essay is intertwined with real science (information on dark matter and the possible geometries of the universe) in such a seamless manner that the average reader might have difficulty discerning fact from fiction.

Fruitful collaboration can also occur between screenwriters/novelists and the scientists themselves, resulting in fictional science that is as realistic as possible (of course, as the plot allows). An excellent resource is the Science and Entertainment Exchange program of the National Academy of Sciences that brings together writers and scientists for fact checking and consultation. The Exchange's stated goal is to "use the vehicle of popular entertainment media to deliver sometimes subtle, but nevertheless powerful, messages about science" [12]. The Exchange also encourages scientists to work with directors to produce science-based "Behind the Scenes" material for DVDs and Blu-ray packaging, understanding well the large potential audience for such outreach [13]. In working together, Hollywood and the science establishment can counter the public's negative stereotypes concerning science and scientists, hopefully reducing unnecessary fears and simple ignorance while providing the all-important spark of awe that will help recruit the next (hopefully far more diverse) generation of scientists. As David Kirby argues, while we should not expect Hollywood to take over the job of educating the public about science, simply ceasing to promulgate negative stereotypes (or, even better, projecting a more accurate image of science) is an excellent first step [14]. The scientific establishment can take it from there.

Finally, we must all remember that communication is a two-way street. It is supremely egotistical for scientists to assume that information only flows in one direction, that while the general public has plenty to learn *from* us, they have nothing of merit to share *with* us. On the contrary, the Committee on the Science of Science Communication of the National Academy of Science opines that scientists have "an equal duty to listen to the public so as to strengthen the quality of public discourse and increase the perceived and actual relevance of science to society. .... It can also clarify what information society needs and wants from scientists" [15]. Depictions of scientists in the media help to shape the public's perception of science and scientists, for good or evil. The converse is also true—the media uses well-known stereotypes of scientists exactly because they are recognizable to the public. Around and around the collider ring we go. My sincerest hope is that the lessons we have learned about effective communication—whether through scientific papers or safety reports, blogs or song parodies, TV series or novels—will break the cycle of mistrust and mystery surrounding particle physics before the next conspiracy theory or Internet hysteria erupts.

The Outer Limits episode "Production and Decay of Strange Particles" concludes with the narrator predicting that "As man explores the secrets of the universe, strange and inscrutable powers await him, and whether these powers are to become forces of destruction or forces of construction will ultimately depend upon simple but profound human qualities: inspiration, integrity, courage" [16]. Let us continue to be inspired, but let us also inspire the spirit of exploration and wonder in others. Let us continue to embody the highest standards of integrity, freely sharing our knowledge with the entirety of humanity with honesty and transparency. But most importantly, we must remember to be brave, embracing the courage to meet the public halfway. We need to step out of the comfort zone of our professional circle, just as we ask the nonscientist to be open to revising what are sometimes deeply held misconceptions about our discipline. In the immortal words of Carl Sagan, "we are a way for the cosmos to know itself" [17]. It is only right that all of us— scientists and nonscientists in equal measure—take the time to get to know each other better. Let us collide in a meeting of the minds, not as matter and antimatter, but as photons, particles of light. Together we will illuminate the dark corners of society where mistrust and misconception have collected for far too long. This is how we will eventually cure the scourge of particle panic.

References

1. R.J. Sawyer, *Flashforward* (TOR, New York, 1999), pp. 112–113
2. J. Ellis, CERN Colloquium: The LHC is Safe, CERN, http://cds.cern.ch/record/1120625?ln=en

3. R. Howes, Communicating Physics to the Public Is a Valuable Skill, APS News **5**(1) (1996), https://www.aps.org/publications/apsnews/199601/valuable-skills.cfm

4. National Science Board, *Science & Engineering Indicators 2004* (National Science Foundation, Alexandria, 2004), p. 7.4

5. C. Woolston, University Tenure Decisions Still Gloss over Scientists' Public Outreach, Nature, https://www.nature.com/articles/d41586-018-06906-z

6. M.I. Feliú-Mójer, Effective Communication, Better Science, Scientific American, https://blogs.scientificamerican.com/guest-blog/effective-communication-better-science/

7. T.H. Huxley, *Discourses: Biological and Geological Essays* (Appleton & Co, New York, 1898), pp. vii–viii

8. R. Cooter, S. Pumphrey, Separate Spheres and Public Places: Reflections on the History of Science Popularisation and Science in Popular Culture. Hist. Sci. **32**, 237–267 (1994)

9. K. Gavroglu, Science Popularization, Hegemonic Ideology and Commercialized Science. J. Hist. Sci. Technol. **6**, 85–99 (2012)

10. M.L. Huggins, Agnes Mary Clerke. Astrophys. J. **25**, 227 (1907)

11. K. Thorne, *The Science of Interstellar* (W.W. Norton, New York, 2015), p. 2

12. About the Science & Entertainment Exchange, The Science & Entertainment Exchange, http://scienceandentertainmentexchange.org/about/

13. D.A. Kirby, *Lab Coats in Hollywood: Science, Scientists, and Cinema* (MIT Press, Cambridge, 2011), p. 225

14. D.A. Kirby, *Lab Coats in Hollywood: Science, Scientists, and Cinema* (MIT Press, Cambridge, 2011), p. 117

15. National Academies of Sciences, Engineering, and Medicine, *Communicating Science Effectively: A Research Agenda* (National Academies Press, Washington, DC, 2017), p. 19

16. L. Stevens (script), Production and Decay of Strange Particles, The Outer Limits, season 1 (1964)

17. C. Sagan (script), The Shores of the Cosmic Ocean, Cosmos, season 1 (1980)

Appendix: Media Works

Novels, Graphic Novels, and Short Stories

Title	Year	Chapter(s)
A Hole in Texas, Herman Wouk	2004	Introduction, 1, 2, 4, 8
A Wrinkle in Time, Madeleine L'Engle	1962	2
Angels & Demons, Dan Brown	2000	1, 3, 4, 5, 6, 8
Cat's Cradle, Kurt Vonnegut	1963	2, 5, 6, 8
Chaos and Order, Stephen R. Donaldson	1994	4
Cloak and Dagger, Christie Golden	2000	6
Cosm, Gregory Benford	1998	2, 3, 4, 5, 7, Conclusion
Dark Matter, Garfield Reeves-Steven	1990	2, 5, 7
Earth, David Brin	2005	3, 4, 5, 7
Einstein's Bridge, John Cramer	1997	1, 2, 3, 5, 6, 7, Conclusion
Flashforward, Robert J. Sawyer	1999	1, 2, 3, 4, 5, 6, 7, Conclusion
Foundation and Chaos, Greg Bear	1998	6
Frankenstein, Mary Shelley	1818	Introduction, 5
God's Spark, Norman P. Johnson	2012	1, 2, 3, 4
His Dark Materials, Phillip Pullman	1995–2000	Introduction, 6, 7
"How We Lost the Moon", Paul McAuley	1999	5
Hyperion, Dan Simmons	1989	4, 7
Into the Looking Glass, John Ringo	2005	3, 6, 7
Jurassic Park, Michael Crichton	1990	2, 5
Krone Ascending, J. Craig Wheeler	2012	3
Life, the Universe and Everything, Douglas Adams	1982	5
Luna Marine, Ian Douglas	1990	4
Olympos, Dan Simmons	2005	3

(continued)

© Springer Nature Switzerland AG 2019
K. Larsen, *Particle Panic!*, Science and Fiction,
https://doi.org/10.1007/978-3-030-12206-5

Title	Year	Chapter(s)
Ring, Stephen Baxter	1994	1, 6
Shadow of Heaven, Christie Golden	2000	6, Conclusion
Star Bright, Martin Caidin	1980	2, 3, 4, 5
"The Borderland of Sol", Larry Niven	1975	7
The Breach/Ghost Country/Deep Sky, Patrick Lee	2010–2	4, 7
"The Collider", James M.M. Baldwin	2012	4
The Doomsday Effect, Thomas Wren	1986	3, 7
The Forge of God, Greg Bear	1987	6
The God Particle, Richard Cox	2005	2, 6, Conclusion
"The Hole Man", Larry Niven	1974	7
The Krone Experiment J. Craig Wheeler	1986	2, 3, 5, 7
The Terror at CERN, Franklin Clermont	2015	1, 4
The Voices at CERN, Franklin Clermont	2014	1, 4, 5, 7
Timescape, Gregory Benford	1980	4, 7, 8
Transformers Timeline: Generation 2: Redux	2010	1

Films

Title	Year	Chapter(s)
12:01	1993	5
2012	2009	3, 6
Angels & Demons	2009	2, 5, 6, 8
Annihilation Earth	2009	5, 6
Dark Storm	2006	6
Decay	2012	Introduction, 2, 5, 6, 8
End:Day	2005	3
Futurama: Bender's Game	2008	4, 6
Ghostbusters	1984	3, 4, 5
GI Joe: The Rise of Cobra	2009	3, 4
Harlock: Space Pirate	2013	6
Heroes (Helden)	2013	2, 3
Iron Man 2	2010	1
Prince of Darkness	1987	4, 7
Rift	2009	7
Solaris	2002	6
Spider-Man 3	2007	5
Star Trek	2009	7
Terminator 3: The Rise of the Machine	2003	1
The Big Bang	2010	4
The Black Hole	2006	5, 7
The Cloverfield Paradox	2018	1, 7
The Krone Experiment	2005	2, 3
The Void	2001	1, 2, 3, 4, 5, 7, 8
Thor: The Dark World	2013	6, 7
Xtro II	1990	2, 4, 7

TV or Web Series

Title	Year	Chapter(s)
Agent Carter	2015–16	6
Agents of S.H.I.E.L.D.	2013–	5, 6, 7
American Dad!	2005–	5
Darker than Black: Gemini of the Meteor	2007–10	7
Eve of Destruction	2013	6
Exosquad	1993–4	6
FlashForward	2009–10	1, 2, 4, 7
Futurama	1999–2013	Introduction, 6
Lab Rats	2012–16	7
Lexx	1997–2002	Introduction, 6
Lost	2004–10	3, 4, 5
Odyssey 5	2002–4	6
Rick and Morty	2013–	6
Sanctuary	2008–11	Introduction
South Park	1997–	5
Star Trek (all series)	1966–	1, 6, 7
Terra Nova	2011	7
The Adventures of the Galaxy Rangers	1986–9	7
The Apocalypse Diaries	2016	4, 5
The Big Bang Theory	2007–	1
The Boat (El Barco)	2011–13	1, 2, 8
The Flash	2014–	1, 2, 3, 4, 5, 6, 7
The Invisible Man	2000–2	4, 5
The Outer Limits	1963–5; 1995–2002	1, 3, 4, Conclusion
The Simpsons	1989–	7
The Sparticle Mystery	2011–15	4, 5
The X-Files	1993–2002	4, 5
Warehouse 13	2009–14	5

Index

© Springer Nature Switzerland AG 2019
K. Larsen, *Particle Panic!*, Science and Fiction,
https://doi.org/10.1007/978-3-030-12206-5

Printed in the United States
By Bookmasters